W9-ABM-716

STRUCTURED HEREDITARY SYSTEMS

MONOGRAPHS AND TEXTBOOKS IN PURE AND APPLIED MATHEMATICS

30. *J. S. Golan,* Localization of Noncommutative Rings (1975)
31. *G. Klambauer,* Mathematical Analysis (1975)
32. *M. K. Agoston,* Algebraic Topology: A First Course (1976)
33. *K. R. Goodearl,* Ring Theory: Nonsingular Rings and Modules (1976)
34. *L. E. Mansfield,* Linear Algebra with Geometric Applications: Selected Topics (1976)
35. *N. J. Pullman,* Matrix Theory and Its Applications (1976)
36. *B. R. McDonald,* Geometric Algebra Over Local Rings (1976)
37. *C. W. Groetsch,* Generalized Inverses of Linear Operators: Representation and Approximation (1977)
38. *J. E. Kuczkowski and J. L. Gersting,* Abstract Algebra: A First Look (1977)
39. *C. O. Christenson and W. L. Voxman,* Aspects of Topology (1977)
40. *M. Nagata,* Field Theory (1977)
41. *R. L. Long,* Algebraic Number Theory (1977)
42. *W. F. Pfeffer,* Integrals and Measures (1977)
43. *R. L. Wheeden and A. Zygmund,* Measure and Integral: An Introduction to Real Analysis (1977)
44. *J. H. Curtiss,* Introduction to Functions of a Complex Variable (1978)
45. *K. Hrbacek and T. Jech,* Introduction to Set Theory (1978)
46. *W. S. Massey,* Homology and Cohomology Theory (1978)
47. *M. Marcus,* Introduction to Modern Algebra (1978)
48. *E. C. Young,* Vector and Tensor Analysis (1978)
49. *S. B. Nadler, Jr.,* Hyperspaces of Sets (1978)
50. *S. K. Segal,* Topics in Group Rings (1978)
51. *A. C. M. van Rooij,* Non-Archimedean Functional Analysis (1978)
52. *L. Corwin and R. Szczarba,* Calculus in Vector Spaces (1979)
53. *C. Sadosky,* Interpolation of Operators and Singular Integrals: An Introduction to Harmonic Analysis (1979)
54. *J. Cronin,* Differential Equations: Introduction and Quantitative Theory (1980)
55. *C. W. Groetsch,* Elements of Applicable Functional Analysis (1980)
56. *I. Vaisman,* Foundations of Three-Dimensional Euclidean Geometry (1980)
57. *H. I. Freedman,* Deterministic Mathematical Models in Population Ecology (1980)
58. *S. B. Chae,* Lebesgue Integration (1980)
59. *C. S. Rees, S. M. Shah, and C. V. Stanojević,* Theory and Applications of Fourier Analysis (1981)
60. *L. Nachbin,* Introduction to Functional Analysis: Banach Spaces and Differential Calculus (R. M. Aron, translator) (1981)
61. *G. Orzech and M. Orzech,* Plane Algebraic Curves: An Introduction Via Valuations (1981)
62. *R. Johnsonbaugh and W. E. Pfaffenberger,* Foundations of Mathematical Analysis (1981)
63. *W. L. Voxman and R. H. Goetschel,* Advanced Calculus: An Introduction to Modern Analysis (1981)
64. *L. J. Corwin and R. H. Szcarba,* Multivariable Calculus (1982)
65. *V. I. Istrătescu,* Introduction to Linear Operator Theory (1981)
66. *R. D. Järvinen,* Finite and Infinite Dimensional Linear Spaces: A Comparative Study in Algebraic and Analytic Settings (1981)

67. *J. K. Beem and P. E. Ehrlich,* Global Lorentzian Geometry (1981)
68. *D. L. Armacost,* The Structure of Locally Compact Abelian Groups (1981)
69. *J. W. Brewer and M. K. Smith, eds.,* Emmy Noether: A Tribute to Her Life and Work (1981)
70. *K. H. Kim,* Boolean Matrix Theory and Applications (1982)
71. *T. W. Wieting,* The Mathematical Theory of Chromatic Plane Ornaments (1982)
72. *D. B. Gauld,* Differential Topology: An Introduction (1982)
73. *R. L. Faber,* Foundations of Euclidean and Non-Euclidean Geometry (1983)
74. *M. Carmeli,* Statistical Theory and Random Matrices (1983)
75. *J. H. Carruth, J. A. Hildebrant, and R. J. Koch,* The Theory of Topological Semigroups (1983)
76. *R. L. Faber,* Differential Geometry and Relativity Theory: An Introduction (1983)
77. *S. Barnett,* Polynomials and Linear Control Systems (1983)
78. *G. Karpilovsky,* Commutative Group Algebras (1983)
79. *F. Van Oystaeyen and A. Verschoren,* Relative Invariants of Rings: The Commutative Theory (1983)
80. *I. Vaisman,* A First Course in Differential Geometry (1984)
81. *G. W. Swan,* Applications of Optimal Control Theory in Biomedicine (1984)
82. *T. Petrie and J. D. Randall,* Transformation Groups on Manifolds (1984)
83. *K. Goebel and S. Reich,* Uniform Convexity, Hyperbolic Geometry, and Nonexpansive Mappings (1984)
84. *T. Albu and C. Năstăsescu,* Relative Finiteness in Module Theory (1984)
85. *K. Hrbacek and T. Jech,* Introduction to Set Theory, Second Edition, Revised and Expanded (1984)
86. *F. Van Oystaeyen and A. Verschoren,* Relative Invariants of Rings: The Noncommutative Theory (1984)
87. *B. R. McDonald,* Linear Algebra Over Commutative Rings (1984)
88. *M. Namba,* Geometry of Projective Algebraic Curves (1984)
89. *G. F. Webb,* Theory of Nonlinear Age-Dependent Population Dynamics (1985)
90. *M. R. Bremner, R. V. Moody, and J. Patera,* Tables of Dominant Weight Multiplicities for Representations of Simple Lie Algebras (1985)
91. *A. E. Fekete,* Real Linear Algebra (1985)
92. *S. B. Chae,* Holomorphy and Calculus in Normed Spaces (1985)
93. *A. J. Jerri,* Introduction to Integral Equations with Applications (1985)
94. *G. Karpilovsky,* Projective Representations of Finite Groups (1985)
95. *L. Narici and E. Beckenstein,* Topological Vector Spaces (1985)
96. *J. Weeks,* The Shape of Space: How to Visualize Surfaces and Three-Dimensional Manifolds (1985)
97. *P. R. Gribik and K. O. Kortanek,* Extremal Methods of Operations Research (1985)
98. *J.-A. Chao and W. A. Woyczynski, eds.,* Probability Theory and Harmonic Analysis (1986)
99. *G. D. Crown, M. H. Fenrick, and R. J. Valenza,* Abstract Algebra (1986)
100. *J. H. Carruth, J. A. Hildebrant, and R. J. Koch,* The Theory of Topological Semigroups, Volume 2 (1986)

101. *R. S. Doran and V. A. Belfi,* Characterizations of C*-Algebras: The Gelfand-Naimark Theorems (1986)
102. *M. W. Jeter,* Mathematical Programming: An Introduction to Optimization (1986)
103. *M. Altman*, A Unified Theory of Nonlinear Operator and Evolution Equations with Applications: A New Approach to Nonlinear Partial Differential Equations (1986)
104. *A. Verschoren*, Relative Invariants of Sheaves (1987)
105. *R. A. Usmani,* Applied Linear Algebra (1987)
106. *P. Blass and J. Lang,* Zariski Surfaces and Differential Equations in Characteristic $p > 0$ (1987)
107. *J. A. Reneke, R. E. Fennell, and R. B. Minton*. Structured Hereditary Systems (1987)
108. *H. Busemann and B. B. Phadke*, Spaces with Distinguished Geodesics (1987)
109. *R. Harte*, Invertibility and Singularity for Bounded Linear Operators (1987)

Other Volumes in Preparation

STRUCTURED HEREDITARY SYSTEMS

James A. Reneke
Robert E. Fennell
Clemson University
Clemson, South Carolina

Roland B. Minton
Roanoke College
Salem, Virginia

MARCEL DEKKER, INC. New York and Basel

Library of Congress Cataloging in Publication Data

Reneke, James A., [date]
Structured hereditary systems.

(Monographs and textbooks in pure and applied mathematics ; 107)
Includes bibliographies and index.
1. Control theory. 2. Kernel functions.
3. Hilbert space. 4. Mathematical optimization.
5. Stochastic processes. I. Fennell, Robert E.
II. Minton, Roland B., [date] III. Title.
IV. Title: Hereditary systems. V. Series: Monographs and textbooks in pure and applied mathematics ; v. 107.
QA402.3.R46 1987 629.8'312 87-5066
ISBN 0-8247-7772-7

MARCEL DEKKER, INC.

270 Madison Avenue, New York, New York, 10016

Current printing (last digit)
10 9 8 7 6 5 4 3 2 1

PRINTED IN THE UNITED STATES OF AMERICA

Preface

Problems of distributed control, computation, and sensing in large scale systems represent the future direction of systems theory and applications. Compartmental models are used by biologists to understand mechanisms present in a complex ecosystem. Environmental regulation presents policy makers with control problems for multicomponent systems with conflicting interests and authority. The distributed control of large space structures poses significant modeling, design, and analysis problems for the control engineer. Problems of restricted decision making and communication arise naturally in military C^3 applications. Our objective in these notes is to present an abstract framework useful for clarifying some of these large scale system concepts and problems.

Within a Hilbert space setting, we present methods for analysis and control of linear hereditary systems. Such systems arise naturally in the description of biological, physical, and mechanical processes. Viewing system variables (signals) as time varying functions leads to two equivalent operator descriptions of linear hereditary systems. Operators from one class provide an input/output description while operators from the other class provide a local description of the system dynamics. Restrictions placed upon the underlying function spaces allow us to describe hereditary systems by operator equations defined on a reproducing kernel Hilbert space. The structure of such spaces facilitates the description of system concepts and the solution of optimization and approximation problems. Our approach is based upon integral equations rather than differential equations and does not require state space descriptions of system dynamics. This approach may be contrasted with semigroup or abstract evolution equation descriptions.

Algebraic relationships between operators describing system components allow us to present a graphical description of *structured hereditary systems*. In terms of signal diagrams, one can think of putting together components to form more complex systems or decomposing a system into simpler components.

In Chapter II we take up optimal control of hereditary systems. Each problem is posed as a constrained optimization problem. Such problems in a reproducing kernel Hilbert space setting are solved using the necessary conditions from the Lagrange Multiplier Theorem. Technical difficulties that arise from not forcing a state space formulation of the dynamics or allowing delays in the input and observation processes are managed by requiring the cost functional to be of a special form. This approach permits us to obtain representations of all the operators appearing in the necessary conditions and hence explicit representations of the optimal control.

In Chapter III a class of polygonal functions is introduced and projection methods are used to obtain corresponding finite dimensional operator approximations. These polygons arise naturally in a reproducing kernel Hilbert space. Convergence results are presented and the discrete structure of operator approximations is clarified. Within this setting, we analyze system identification, parameter estimation, and the numerical solution of optimal control problems.

In Chapter IV we show that stochastic systems may be studied within the same general framework . Background information on stochastic differential equations is presented. Control problems are solved for state space and hereditary systems with and without state dependent noise.

In Chapter V we return to the problem of control for deterministic systems. All of the previous results apply to systems with a central controller. In this chapter, we introduce control problems for large scale systems, i.e., systems with two or more somewhat autonomous controllers. Because of restrictions on the flow of information between controllers, the goal becomes system coordination rather than system optimization.

The material is mostly self-contained. A summary of necessary Hilbert space properties is presented in Chapter I. Background on optimization theory and the Lagrange Multiplier Theorem in a Hilbert space setting is included in Chapter II. The review of stochastic hereditary systems presented in Chapter IV is based on principles of measure theory. An introductory course on linear analysis is a prerequisite for these notes. A course on measure and integration theory is prerequisite for Chapter IV. Chapters II, III, IV, and V are, for the most part, self contained; each chapter relies on the introductory material presented in Chapter I.

At this point, we acknowledge the work of two former Clemson University graduate students, S. L. Benz and S. B. Black. During the preparation of this manuscript support has been received from the NASA Langley Research Center, Naval Coastal Systems Center, the Air Force Office of Scientific Research, and the Institute for Computer Applications in Science and Engineering. We greatly appreciate this support. Finally, we wish to thank Ms. Dianne Myers and Ms. Kelly Wood for typing the entire manuscript.

James A. Reneke
Robert E. Fennell
Roland B. Minton

Contents

I
Introduction to Structured Systems

1. Purpose and Scope

Modern technology is characterized by increased automation and the design of larger and more distributed systems. Given these ambitions, it is vital to develop a mathematical theory which applies to large scale systems as well as one component systems. Such a theory should enable us to analyze stochastic systems as well as deterministic systems, allow for hereditary system dynamics, and readily admit efficient numerical algorithms. Finally, this theory should be accessible to interested researchers in all disciplines.

We will present a general framework for the analysis of linear hereditary systems, an abstract framework which is useful for clarifying basic system concepts and problems. Approaching structured hereditary systems as collections of interacting components, we will take up system problems of control and identification for both deterministic and stochastic systems using reproducing kernel Hilbert space methods [1, 10]. The notes are mostly self contained requiring only basic concepts of linear analysis.

Outline of notes. It is common to view a system as an input/output relation in which the input and output variables (signals) are functions of time, and the input/output relation is described by an operator equation defined on an appropriate function space [2, 3, 4]. Our approach in the remainder of this chapter requires that system input, output, and response functions belong to an underlying function space whose topology is induced by a family of seminorms. Within this setting two classes of linear operators are introduced.

Operators from these classes yield two equivalent realizations of the system dynamics. Intuitively, operators from one class provide an input/output description, while operators from the other class provide a local description of the system dynamics. A further assumption on system functions allows us to describe linear hereditary systems by operator equations defined on a reproducing kernel Hilbert space. The structure of such spaces facilitates the analysis of system problems.

Typical examples of the kinds of dynamics to be considered include

1. Standard finite dimensional control systems of the form $x' = \alpha x + \beta u + v$, where α and β denote $n \times n$ and $n \times m$ matrices and x, u, v denote state, control and disturbance variables of appropriate dimensions.
2. Volterra integral equations of the form $x(t) = f(t) + \int_{[0,t]} g(t, s)x(s)\, ds$.
3. Retarded functional differential equations [6] of the form $x'(t) = g(t, x_t) + \beta u(t) + v(t)$ where g is linear in the second argument and $x_t(\theta) = x(t + \theta)$ for $-r \leq \theta \leq 0$.
4. Stochastic differential or integral equations [11] of the form (1) - (3), with f, g and v random functions.

Our abstraction of such equations is based upon an integral equation rather than differential or abstract evolution equation description of system dynamics. Systems will be described by operator equations of the form

$$h = Af \quad \text{or} \quad h = f + Bh$$

where A and B denote operators defined on appropriate function spaces and f, h denote input and output variables. Within this framework, the first equation is viewed as the global or input-output description of the system dynamics while the second equation provides a local description of the dynamics.

In this first chapter, we are concerned with the internal part of a system, not with the description of methods for internalizing inputs or measuring system outputs. We will introduce the notion of a structured hereditary system using these basic building blocks, i.e., the A's and B's. The systems discussed in this introductory chapter will be elaborated in later developments.

In Chapter II we take up optimal control of hereditary systems. We pose each problem as a constrained optimization problem [9]. Such problems in a reproducing kernel Hilbert space setting are solved using the necessary conditions that come from the Lagrange Multiplier Theorem. We add significantly to the realism of system control problems by adding the elements of an input process and an observation process. Technical difficulties that arise from not forcing a state space formulation of the dynamics or allowing delays in the input and observation processes are managed by requiring the cost functional to be of a special form. This approach permits us to obtain representations of all the operators appearing in the necessary conditions and hence an explicit representation of the optimal control.

In Chapter III a class of polygonal functions is introduced along with corresponding operator approximations. These polygonal functions arise naturally in a reproducing kernel Hilbert space setting. Convergence results are presented and attention is given to the numerical implementation of the methods. Within this context, system identification, parameter estimation, and optimal control problems are analyzed.

In Chapter IV we solve a class of control problems using the same techniques as in the second chapter. We introduce some concepts from time series analysis to compute the output covariance matrix [13], which is basic to the solution of various filtering and parameter estimation problems. To analyze a system with state dependent noise (e.g., equation (2) with g random), we extend the class of operators considered and solve a simplified control problem.

In Chapter V we return to problems of control for deterministic systems. All of the previous results apply to systems with a central controller. In Chapter V we introduce control problems for large scale systems, i.e., systems with two or more somewhat autonomous controllers. Because of restrictions on the flow of information between controllers, the goal becomes system coordination rather than system optimization.

2. The Basic Building Blocks of Structured Systems

Fundamental to modeling a complex physical system is a decomposition of the system into manageable units or components. Having successfully modelled the components, the overall system model is obtained by joining together the component models. Thus structured systems, systems composed of components joined in recognizable ways, can arise from immediate considerations. However, there are also some conceptual benefits from thinking of systems as made up of components such as the plant, controller, and observer in control theory [15]. For instance, we are prompted to ask meaningful questions about component interactions and the relation of these interactions to overall system performance.

In this section we will introduce the basic component models and a method for connecting the components into an overall system model, i.e., a structured system. We begin our formal study of hereditary systems by introducing two classes of linear operators which will serve as descriptors for system dynamics. Having presented in Section 1 a sampling of the variety of systems which can be classified as hereditary systems and of some problems for those systems, we want to step back and treat the notion of hereditary system from a more abstract viewpoint. Our objective is a broader per- spective on the one hand and a refinement of our solution techniques on the other.

The classes of operators. We will assume for the remainder of these notes that S is a number interval containing 0, $\{X, |\cdot|\}$ is a Banach space, and G is a linear space of functions from S into X. Furthermore, we will assume that N is a function from S^+, the nonnegative part of S, into the class of all pseudonorms on G such that

1. For w in S^+ and f in G, $N_w(f) = 0$ if and only if $f(u) = 0$ for all u in S such that $u \leq w$.
2. For $0 \leq u \leq v$ in S^+ and f in G, $N_u(f) \leq N_v(f)$.
3. $\{G, N\}$ is complete, i.e., if the sequence $\{f_n\}_{n=0,\infty}$ in G is Cauchy with respect to N_u for each u in S^+ then there is a g in G such that $\{f_n\}$ has limit g with respect to N_u for each u in S^+.

Let $\boldsymbol{H}$ denote the class of linear functions from G into G, and suppose that P is a function from S^+ into the idempotent members of $\boldsymbol{H}$ such that if each of u and v is in S^+ and f is in G then $N_v(P_u f) = N_w(f)$, where $w = \min(u, v)$.

An example of the underlying function space will consist of a four tuple {S, G, N, P}. We will concentrate on intervals S of the form [-r, T], where r is non-negative and T is either a positive number or infinity. In the standard examples we take G to be either the continuous, quasicontinuous, or absolutely continuous functions, and

$$[P_u f](x) = \begin{cases} f(x) & x \le u \\ f(u) & u \le x \end{cases}$$

for f in G and (u, x) in $S \times S$. For either the continuous or quasicontinuous case let $N_u(f) = \sup_{x \le u} | f(x) |$. In the absolutely continuous case, we will use the variational norm $N_u(f) = | f(-r) | + \int_{[-r,u]} | df |$.

Other useful examples will be constructed later from subspaces of the standard examples. However, we are not limited to this type of example. Let $G = L_p(S)$ for $1 \le p < \infty$ and $N_u(f) = \{\int_{[l,u]} | f |^p dI\}^{1/p}$ for f in G, and u in S. Here $\boldsymbol{l}$ is the left boundary of S. For this case let

$$[P_u f](x) = \begin{cases} f(x) & x \le u \\ 0 & u \le x \end{cases}$$

Before introducing the classes of operators used to describe basic system components, let us review briefly the left and right Stieltjes integrals $(L)\int_{[a,b]} m\, dk$ and $(R)\int_{[a,b]} m\, dk$ for nondecreasing real functions m and k on a number interval [a, b]. The left integral is the limit through refinements of left sums $\Sigma_{p=1,n}\, m(s_{p-1})\cdot(k(s_p) - k(s_{p-1}))$, where $\{s_p\}_{p=0,n}$ partitions [a, b]. Note that if $u \le v \le w$ then $m(u)(k(w) - k(u)) \le m(u)(k(v) - k(u)) + m(v)(k(w) - k(v))$. Hence $(L)\Sigma_s\, m\, dk \le (L)\Sigma_t\, m\, dk$ for refinements t of s. The notation $dk(u, v) = k(v) - k(u)$ will be used frequently throughout these notes. Further, $(L)\Sigma_s\, m\, dk < m(b)(k(b) - k(a))$, i.e., the approximating sums are bounded above. Thus $(L)\int_{[a,b]} m\, dk$ is the least upper bound of the set of all left sums $(L)\Sigma_s\, m\, dk$ based on partitions s of [a, b]. In a similar manner, $(R)\int_{[a,b]} m\, dk$ is the greatest lower bound of the set of all right sums $(R)\Sigma_s\, m\, dk = \Sigma_{p=1,n}\, m(s_p)(k(s_p) - k(s_{p-1}))$ based on partitions $\{s_p\}_{p=0,n}$ of [a, b].

<u>Exercise</u>. $(L)\int_{[a,b]} m\, dk = m(b)k(b) - m(a)k(a) - (R)\int_{[a,b]} m\, dk$. Therefore $(L)\int_{[a,b]} m\, dm \le (m^2(b) - m^2(a))/2$.

<u>Exercise</u>. For each positive integer p, $(L)\int_{[a,b]} (m(x) - m(a))^p\, dm(x) \le (m(b) - m(a))^{p+1}/(p+1)$.

The classes of operators used to describe basic system components will now be defined. Let $\boldsymbol{B}$ denote the subset of $\boldsymbol{H}$ to which B belongs only in case

1. $[Bf](u) = 0$ for each f in G and u in S which does not exceed 0.
2. For each positive number w in S^+ there is a nondecreasing function k from S to the numbers such that

$$N_w(P_v Bf - P_u Bf) \leq (R) \int_u^v N_x(f)\, dk(x)$$

for each f in G and subinterval [u, v] of [0, w].

Given that (B, k) is a pair satisfying (2) on [0, w] we will say that k is a variation for B on [0, w]. Note that B is a continuous linear transformation of {G, N}.

Exercise. In case G is the space of continuous functions, $[P_u f](x) = f(\min(u, x))$, and $N_u(f) = \sup_{x \leq u} | f(x) |$, show that (2) is equivalent to

2′. For each positive number w in S there is a nondecreasing function k from S to the numbers such that

$$| [Bf](v) - [Bf](u) | \leq \int_u^v N_x(f)\, dk(x)$$

for each f in G and subinterval [u, v] of [0, w].

Exercise. Suppose that $S = S^+$, G is the collection of all quasicontinuous functions from S into X, $N_x(f) = \sup_{u \leq x} | f(u) |$ for each f in G and x in S, and

$$[P_x f](u) = \begin{cases} f(u) & 0 \leq u \leq x \\ 0 & x < u \end{cases}$$

for each f in G and (x, u) in $S \times S$. Suppose that g is a function from S into the continuous linear transformations of X and k is a nondecreasing function from S into the numbers such that $| dg(u, v)x | \leq dk(u, v) | x |$ for each subinterval [u, v] of S and x in X. Let B denote the member of $\boldsymbol{H}$ defined by $[Bf](u) = (R) \int_{[0,u]} dg\, f$ for each f in G and u in S. Then B is a member of $\boldsymbol{B}$.

Example 1. Suppose that $S = S^+$, G is the collection of all continuous functions from S into the space X of real numbers with the usual norm, $N_x(f) = \sup_{u \leq x} | f(u) |$ for each f in G and x in S, and

$$[P_x f](u) = \begin{cases} f(u) & 0 \le u \le x \\ f(x) & x < u \end{cases}$$

for each f in G and (x, u) in $S \times S$. Suppose that g is a continuous function from $S \times S$ into X such that g_1 (the partial derivative of g with respect to the first argument) is also continuous. Let B denote the member of ***H*** defined by $[Bf](u) = \int_{[0,u]} g(u, s)f(s)\, ds$, for each f in G and u in S. Then, since $[Bf](u) = \int_{[0,u]} [g(t, t)(f(t) + \int_{[0,t]} g_1(t, s)f(s)\, ds]\, dt$ for each f in G and u in S, B is a member of ***B***.

Example 2. Suppose that r is a positive number, $S = [-r, \infty]$, G is the collection of all absolutely continuous functions from S into X, $N_x(f) = | f(-r) | + \int_{[-r,x]} | df |$ for each f in G and x in S^+, and

$$[P_x f](u) = \begin{cases} f(u) & -r \le u \le x \\ f(x) & x \le u \end{cases}$$

for each f in G and (x, u) in $S^+ \times S$. Let Y denote the space of continuous functions from [-r, 0] into X with supremum norm $| \cdot |_y$, and for each f in G and s in S^+, let f_s be the member of Y defined by $f_s(u) = f(s + u)$ for each u in [-r, 0]. Suppose that Θ is a continuous function from $S^+ \times Y$ into X such that $\Theta(s, \cdot)$ is linear for each s in S^+. Also, suppose m is a continuous nonegative function defined on S such that $| \Theta(s, f) | \le m(s) | f |_Y$ for each s in S^+ and f in Y. Let B be the member of ***H*** defined by

$$[Bf](u) = \begin{cases} 0 & -r \le u \le 0 \\ \int_0^y \Theta(s, f_s)\, ds & 0 \le u \end{cases}$$

for each f in G and u in S. Then B is a member of ***B***.

Theorem 1.1. If B is in ***B*** then I - B is a reversible function from G onto G, where I denotes the identity in ***H***.

Proof. There is a nondecreasing function k on S such that $N_u(Bf) \le (R)\int_{[0,u]} N_x(f)\, dk(x) \le N_u(f) dk(0, u)$ for each f in G and u in S^+. Hence for each u in S^+

$$N_u(B^2f) \le (R)\int_0^u N_x(Bf)\,dk(x)$$
$$\le N_u(f)(R)\int_0^u (k(x) - k(0))\,dk(x)$$
$$\le N_u(f)(dk(0, u))^2/2$$

By induction, for each positive integer p, $N_u(B^pf) \le N_u(f)(dk(0, u))^p/p!$ for each f in G and u in S^+. Therefore $(I - B)f = 0$ implies $N_u(f) \le N_u(f)(dk(0, u))^p/p!$ for each u in S^+ and positive integer p, consequently $f = 0$ and I - B is reversible, i.e., the inverse of I - B is a function.

Also note that $f + \Sigma_{p=1,\infty} B^pf$ converges with respect to N_u for each f in G and u in S^+. Let $A = I + \Sigma_{p=1,\infty} B^p$. If f is in G and h = Af then

$$h = f + \sum_{p=1}^{\infty} B^pf = f + B(f + \sum_{p=2}^{\infty} B^{p-1}f) = f + Bh$$

i.e., (f, h) is in $(I - B)^{-1}$ and so I - B maps G into G. We can also deduce from the inequalities that $N_u(Af) \le N_u(f)\exp(dk(0, u))$ for each f in G and u in S^+. □

The second class of operators used to define basic system properties is defined as follows. Let ***A*** denote the subset of ***H*** to which A belongs only in case A - I is in ***B***. One should note that B in ***B*** implies that $(I - B)^{-1}$ is in ***A***. Specific representations of some operators in ***A*** will be given in examples to follow.

Theorem 1.2. If A is in ***A*** then A is a reversible function from G onto G and $I - A^{-1}$ is in ***B***. Also $[Af](u) = f(u)$ for each f in G and $u \le 0$ in S.

Proof. Since I - A is in ***B***, by the previous theorem A is a reversible function from G onto G. If f is in G and $u \le 0$ then $[Af](u) = f(u) + [(A - I)f)](u) = f(u)$. Hence $[(I - A^{-1})f](u) = f(u) - [A^{-1}f](u) = f(u) - f(u) = 0$ for $u \le 0$. Let B = I - A have variation function k. Then as in the proof of Theorem 1.1, $A^{-1} = (I - B)^{-1}$ and $N_u(A^{-1}f) \le N_u(f)\exp(k(u) - k(0))$ for f in G and u in S^+. Thus

$$N_w(P_v(I - A^{-1})f - P_u(I - A^{-1})f) = N_w(P_v(A - I)A^{-1}f - P_u(A - I)A^{-1}f)$$

$$\leq (R) \int_u^v N_x(A^{-1}f)\, dk(x)$$

$$\leq (R) \int_u^v N_x(f)(\exp(k(x) - k(0)))\, dk(x)$$

$$\leq (R) \int_u^v N_x(f)\, d \exp(k(x) - k(0))$$

and so $I - A^{-1}$ is in $\boldsymbol{B}$. □

Exercise. Verify that $\boldsymbol{A}$ and $\boldsymbol{B}$ are convex sets.

Exercise. A operator D in $\boldsymbol{H}$ is said to be causal if $P_tD = P_tDP_t$ for each t in S. Show that every element of $\boldsymbol{A}$ and $\boldsymbol{B}$ is causal.

Exercise. For $r > 0$ let D be an element of $\boldsymbol{H}$ defined by $[Df](t) = 0$ for $t \leq 0$ and $[Df](t) = f(t - r)$ for $0 \leq t$. Show that D is causal but that there is a t in S such that $P_tD \neq DP_t$.

Exercise. If B is in $\boldsymbol{B}$ with variation function k and $A = (I - B)^{-1}$, then $\exp(k - k(0))$ is a variation function for A - I.

Exercise. (Gronwall's Inequality). If $0 \leq f(t) \leq c\,(R) \int_{[0,t]} f(s)g(s)\, dk(s)$ for $0 \leq t \leq T$, where c is a positive constant and $g > 0$ on $[0, T]$, then for $0 \leq t \leq T$ one obtains $f(t) \leq c \exp((R) \int_{[0,t]} g(s)\, dk(s))$.

Structured systems. A system with dynamics given by $h = f + Bh$, for B in $\boldsymbol{B}$, or $h = Af$, for A in $\boldsymbol{A}$, is called an hereditary system. All of the systems described in the introductory section may be classified as hereditary systems. We can see from the last section that this class of systems is a lot richer than we could indicate with a limited number of examples. In this section we will begin to explore the class of hereditary systems by showing how larger systems can be built up in meaningful ways from elements of $\boldsymbol{A}$ and $\boldsymbol{B}$.

We say that the integral operator $[Bh](t) = \int_{[0,t]} g(t, s)h(s)\, ds$ is formal if the domain of B is unspecified. Since the space G will be in the domain of the formal integral operator, the operator B is the restriction of the formal operator to G. The system $h(t) = f(t) + \int_{[0,t]} g(t, s)\, h(s)\, ds$ for instance, can be represented by $h = f + Bh$, where f and h are from some appropriate function space G. For problems of control and optimization the choices we make for the underlying function spaces play an important role.

Note that although we normally meet the system dynamics in terms of B's, as in the previous paragraph, the dynamics might be given in terms of A's. If $A = (I - B)^{-1}$ then the above system can also be represented by $h = Af$. Laplace transforms provide a pencil and paper method for computing the formal operator $(I - B)^{-1}$ for some examples for which the dimension of X is small. In the next two exercises let X be the space of real numbers, $S = [0, \infty]$, and G be the space of continuous functions from S into X.

Exercise Given $[Bh](t) = \int_{[0,t]} (t - u)h(u)\, du$ use Laplace transforms to show that $[(I - B)^{-1}f](t) = f(t) + (1/2)\int_{[0,t]} (e^{t-u} - e^{u-t})f(u)\, du$.

Exercise Given $[Af](t) = f(t) + \int_{[0,t]} (t - s)f(s)\, ds$ show that $h(t) - [A^{-1}h](t) = \int_{[0,t]} \sin(t - u)h(u)\, du$.

We now describe further algebraic relationships between the classes $\boldsymbol{A}$ and $\boldsymbol{B}$. For $\boldsymbol{H}_1$ and $\boldsymbol{H}_2$ subsets of $\boldsymbol{H}$, let $a\boldsymbol{H}_1 = \{aH \mid H \text{ in } \boldsymbol{H}_1\}$ for each scalar a, $(\boldsymbol{H}_1)^{-1} = \{H^{-1} \mid \text{for reversible } H \text{ in } \boldsymbol{H}_1\}$, $\boldsymbol{H}_1 + \boldsymbol{H}_2 = \{H_1 + H_2 \mid H_1 \text{ in } \boldsymbol{H}_1 \text{ and } H_2 \text{ in } \boldsymbol{H}_2\}$, and $\boldsymbol{H}_1\boldsymbol{H}_2 = \{H_1H_2 \mid H_1 \text{ in } \boldsymbol{H}_1 \text{ and } H_2 \text{ in } \boldsymbol{H}_2\}$.

Theorem 1.3. The following relations hold between $\boldsymbol{A}$ and $\boldsymbol{B}$.

1. $a\boldsymbol{B} = \boldsymbol{B}$ for each nonzero scalar a.
2. $\boldsymbol{B} + \boldsymbol{B} = \boldsymbol{B}$
3. $\boldsymbol{A} + \boldsymbol{B} = \boldsymbol{A}$
4. $\boldsymbol{AB} = \boldsymbol{BA} = \boldsymbol{B}$,
5. $\boldsymbol{A}^{-1} = \boldsymbol{A}$
6. $\boldsymbol{AA} = \boldsymbol{A}$ and $\boldsymbol{BB} \subset \boldsymbol{B}$
7. $\boldsymbol{A} - \boldsymbol{A} = \boldsymbol{B}$

Partial proof. Let us consider just part (4) of the theorem. If A is in $\boldsymbol{A}$, and B is in $\boldsymbol{B}$, and k is a variation for both A - I and B on [0, w], then for $u \leq v \leq w$ in S^+ and f in G

$$\begin{aligned}
N_w(P_vABf - P_uABf) &\leq N_w(P_v(A - I)Bf - P_u(A - I)Bf) + N_w(P_vBf - P_uBf) \\
&\leq (R)\int_u^v N_x(Bf)\,dk(x) + (R)\int_u^v N_x(f)\,dk(x) \\
&\leq (R)\int_u^v N_x(f)(k(x) - k(0))\,dk(x) + (R)\int_u^v N_x(f)\,dk(x) \\
&\leq (R)\int_u^v N_x(f)\,d\{k(x) - k(0))^2/2 + k(x)\} \\
&\leq (R)\int_u^v N_x(f)\,d\exp(k(x) - k(0))
\end{aligned}$$

Also

$$\begin{aligned}
N_x(P_vBAf - P_uBAf) &\leq (R)\int_u^v N_x(Af)\,dk(x) \\
&\leq (R)\int_u^v N_x(f)\exp(k(x) - k(0))\,dk(x) \\
&\leq (R)\int_u^v N_x(f)\,d\exp(k(x) - k(0))
\end{aligned}$$

Thus both AB and BA are in $\boldsymbol{B}$ or both $\boldsymbol{AB}$ and $\boldsymbol{BA}$ are contained in $\boldsymbol{B}$. Since I is in $\boldsymbol{A}$, $\boldsymbol{AB} = \boldsymbol{BA} = \boldsymbol{B}$. □

Exercise. Establish the remaining parts of the theorem.

Exercise. If $P_tP_s = P_{\min(t,s)}$ or $P_tP_s = P_{\max(t,s)}$ for t, s in S and $\boldsymbol{P} = \{P_t \mid t \text{ in } S\}$ then $\boldsymbol{PB} \subset \boldsymbol{B}$.

We can exploit the algebraic properties of the classes $\boldsymbol{A}$ and $\boldsymbol{B}$ by introducing a graphical representation of structured hereditary systems. A directed graph is a finite collection Γ of nodes and arcs such that 1) each arc in Γ contains exactly two nodes, 2) each node in Γ is contained in some arc, and 3) in each arc, one of the nodes is designated the initial node and the other the terminal node [14, 16]. Note

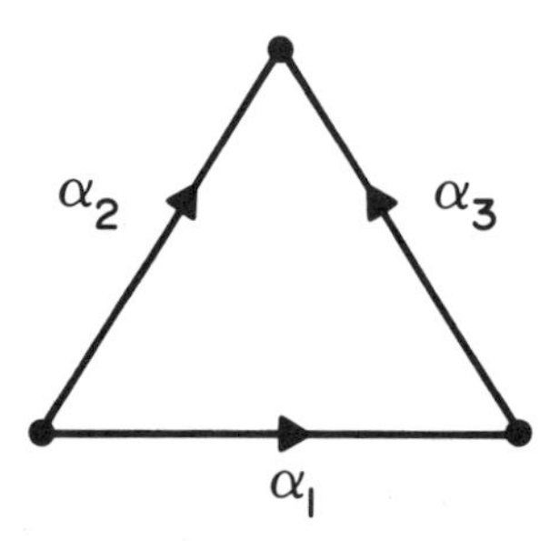

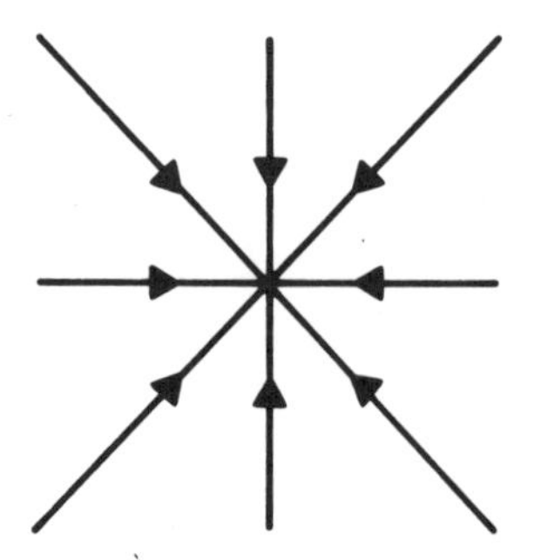

Figure 1.1

that "node" and "arc" are undefined terms. We interpret nodes to be dots and arcs to be line segments. Arrows will be directed from initial nodes to terminal nodes.

A finite sequence $\{\alpha_p\}_{p=0,n}$ of arcs in Γ is called a semipath (path) provided, for p = 1, 2, ..., n, the arcs α_{p-1} and α_p have a node in common (the terminal node of α_{p-1} is the initial node of α_p). For example, in Figure 1.1$\{\alpha_1, \alpha_2, \alpha_3\}$ is a semipath which is not a path. An arc α is said to join an arc ß provided there is a semipath $\{\gamma_p\}_{p=0,n}$ such that $\gamma_0 = \alpha$ and $\gamma_n =$ ß. If α is joined to ß by a path then α is said to precede ß (or ß follows α). If each pair of arcs in Γ are joined then Γ is connected. We only consider connected graphs.

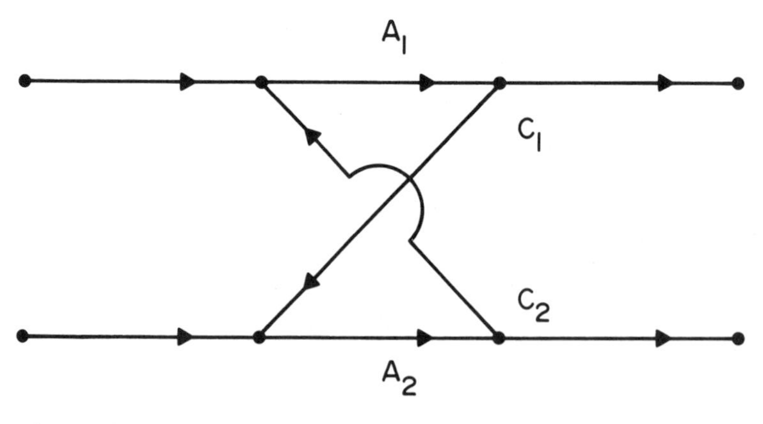

Figure 1.2

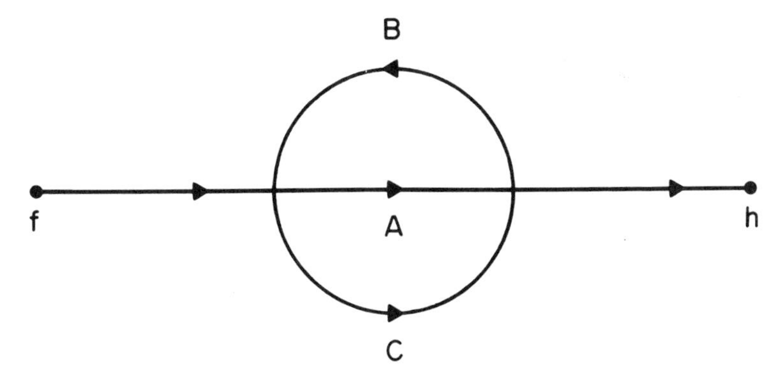

Figure 1.3

If we label the arcs of Γ with members of $\boldsymbol{A} \cup \boldsymbol{B}$ as illustrated in Figure 1.2 (i.e., specify a function from the arcs of Γ into $\boldsymbol{A} \cup \boldsymbol{B}$) then Γ with the labeling is called a diagram. In the illustrations, we will leave the arcs labeled by I blank.

For the moment we restrict our attention to directed graphs Γ such that the following hold.

1. There is only one arc α in Γ with the property that α precedes every other arc in Γ and follows no other arc in Γ. The initial node of α is called the input node.
2. There is only one arc ω in Γ with the property that ω follows every other arc in Γ and precedes no other arc in Γ. The terminal node of ω is called the output node.

An hereditary system with input/output operator A is said to be represented by a signal diagram Γ provided Γ is a diagram (the labeling of the arcs is assumed) and the following hold. For each f in G, if the input node is labeled by f then there is a labeling of remaining nodes by members of G so that the output node is labeled with Af with A in $\boldsymbol{A}$. If h is a node in Γ and Z is the subset of $(\boldsymbol{A} \cup \boldsymbol{B}) \times G$ to which (C, g) belongs provided C is an arc of Γ with initial node g and terminal node h, then $h = \Sigma_{(C,g) \text{ in } Z}\, Cg$. Hereditary systems represented by nondegenerate signal diagrams are called structured hereditary systems.

Theorem 1.4. If A is in $\boldsymbol{A}$ and (B, C) is in $\boldsymbol{B} \times \boldsymbol{B}$ then the system represented by the signal diagram, Figure 1.3, has input/output operator $\{(A + C)^{-1} - B\}^{-1}$.

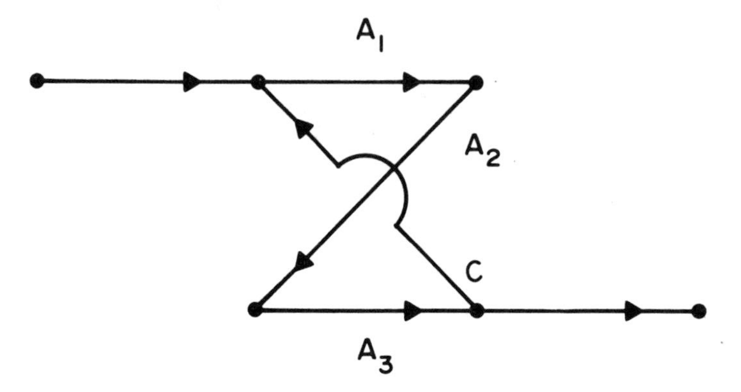

Figure A

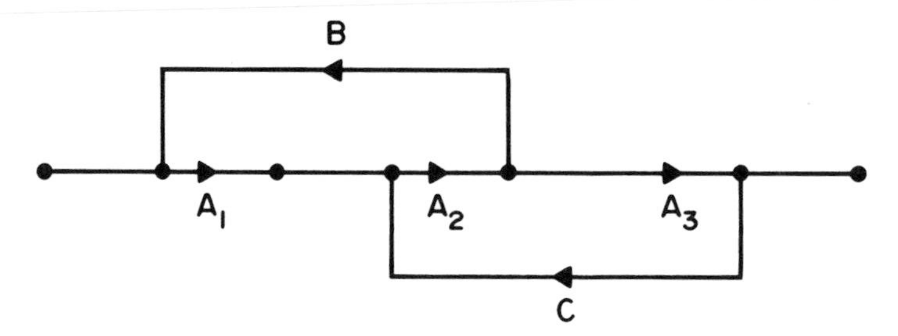

Figure B

Labeling the input node f and the output node h in the above diagram, we have h = A(f + Bh) + C(f + Bh) or $(A + C)^{-1}h = f + Bh$ or $h = ((A + C^{-1} - B)^{-1}f$.

Exercise. Find the input/output operator in Figure A and Figure B, where the A's belong to $\mathcal{A}$ and B, C belong to $\mathcal{B}$.

Problem. Which diagrams among the class of all possible diagrams constructed by following the rules represent hereditary systems?

3. Some Background on Inner Product Spaces

We will outline the basic facts about inner product spaces required for the system techniques to be presented in these notes. Some results will be presented in a form that is convenient for the applications we have in mind; i.e., we will not try to give a general development of all of the concepts [2, 9]. Looking ahead, the system problems that we address have a natural inner product space setting which simplifies the task of representing and ultimately solving for various operators of interest.

Let G_0 be any linear family of real (or complex) valued functions (implicit assumption: each two members of G_0 have the same domain) and suppose that Q_0 is a function from $G_0 \times G_0$ to the real (or complex) numbers such that, for x, y, and z in G_0 and c a number

1. $Q_0(x, y) + Q_0(z, y) = Q_0(x + z, y)$.
2. $Q_0(cx, y) = cQ_0(x, y)$.
3. $Q_0(y, x) = Q_0(x, y)^*$ (complex conjugation).
4. $Q_0(x, x) \geq 0$ and if $Q_0(x, x) = 0$ then $x = 0$.

The ordered pair $\{G_0, Q_0\}$ is called a real (or complex) inner product space. If the distinction between real or complex numbers is unimportant, we will speak of the field of scalars.

Exercise. (Cauchy-Schwarz Inequality) If (x, y) is in $G_0 \times G_0$ then $| Q_0(x, y) |^2 \leq Q_0(x, x)Q_0(y, y)$. Show this assuming, in place of 4), 4′) $Q_0(x, x) \geq 0$.

Exercise. The function N_0 defined on G_0 by $N_0(x) = Q_0(x, x)^{1/2}$ for x in G_0 is a norm for G_0.

If every Cauchy sequence of elements of G_0 converges to a member of G_0, then $\{G_0, Q_0\}$ is said to be complete. Here, the notions of Cauchy sequence and convergence are understood to be with respect to N_0. Complete inner product spaces are also called Hilbert spaces.

Exercise. Suppose that L is a linear function from G_0 to the scalars. L is continuous if and only if there is a number b such that $| L(x) | \leq bN(x)$ for x in G_0. In this case L is referred to as a bounded linear functional.

Exercise. For each y in G_0, $Q_0(\cdot, y)$ is a continuous linear function from G_0 to the scalars. Also Q_0 is a continous function from $G_0 \times G_0$ to the scalars.

Theorem 1.5. (Riesz-Fischer) If $\{G_0, Q_0\}$ is complete and L is a continuous linear function from G_0 to the scalars then there is a member y of G_0 such that $L = Q_0(\cdot, y)$.

Lemma 1.6. Suppose that L is a continuous linear function from G_0 to the scalars and b is the least nonnegative number b′ such that $| L(x) | \leq b'N(x)$ for all x in G_0. If each of x and y is in G_0 then

$$\{b^2Q_0(x, y) - L(y)^*\}^2 \leq \{(v^2N_0^2(x) - | L(x) |^2\}\{b^2N_0^2(y) - | L(y) |^2\}$$

$$bN_0(x - y) - | L(x) - L(y) | \leq (b^2N_0^2(x) - | L(x) |^2)^{1/2} + (b^2N_0^2(y) - | L(y) |^2)^{1/2}$$

Lemma 1.7. If $\{y_p\}_{p=0,\infty}$ is a sequence with values in G_0 such that $N_0(y_p) = 1$, for $p = 0, 1, 2,$ and $\{L(y_p)\}_{p=0,\infty}$ has limit b, then $\{y_p\}_{p=0,\infty}$ converges with respect to N_0.

Theorem 1.8. (Parallelogram Law) If each of x and y is in G_0 then $N_0(x + y)^2 + N_0(x - y)^2 = 2N_0(x)^2 + 2N_0(y)^2$.

Theorem 1.9. (Polar Decomposition) If each of x and y is in G_0 then $Q_0(x, y) = \Sigma_{p=1,4} (i^p/4N_0(x + i^py)^2$, where i is the complex number.

Exercise. Let G_0 denote the set of all sequence $\{x_p\}_{p=0,\infty}$ of complex numbers such that $\Sigma_{p=0,\infty} | x_p |^2$ converges. 1) Show that G_0 is a linear space. Let Q_0 denote the set of ordered pairs to which (u, v) belongs only in case u is in $G_0 \times G_0$ and $v = \Sigma_{p=0,\infty} x_p y_p^*$, where u = (x, y). 2) Show that Q_0 is an inner product for G_0. 3) Show that $\{G_0, Q_0\}$ is complete. Let L denote the function defined on G_0 by $L(x) = \Sigma_{p=0,\infty} x_p/(p + 1)$. 4) Show that L is a continuous linear function from G_0 to the scalars.

We note that we might have started this section by assuming G_0 to be some abstract linear space, rather than a linear space of functions. You will discover however that all of the important examples will be function spaces. Let us assume for the moment that the domain of the elements of G_0 is some fixed set E_0. We will be concerned at the end of this section with linear spaces G of functions from a set E_1 into G_0. Elements of G can be thought of equivalently as functions from $E_0 \times E_1$ to the scalars.

If each of x and y is in G_0 and $Q_0(x, y) = 0$ then x and y are said to be orthogonal. We write $x \perp y$. For a subset M of G_0 the orthogonal complement of M, written $M^\perp$, is the set of vectors in G_0 which are orthogonal to each member of M. The closure of M, written cl(M), is the smallest closed subset of G_0 containing M.

Exercise. If each of x and y is in G_0, $x \neq 0$, and $w = y - (Q_0(y, x)/Q_0(x, x))x$, then $x \perp w$. Here $(Q_0(y, x)/Q_0(x, x))x$ is the projection of y on x.

Theorem 1.10. (Projection Theorem) Suppose that $\{G_0, Q_0\}$ is complete and M is a closed linear subspace of G_0. For each x in G_0 there is a unique member m_0 of M such that $N_0(x - m_0) \leq N_0(x - m)$ for all m in M. Furthermore, if m is in M then $m = m_0$ only in case x - m is orthogonal to M.

Exercise. Suppose that each of x_1 and x_2 is in G_0, x_1 is not a multiple of x_2, and M is the subspace of G_0 spanned by $\{x_1, x_2\}$. Let y be a member of G_0. Find the member of M nearest y.

Theorem 1.11. Suppose that each of M and N is a subset of G_0. Then the following hold:

1) $M^\perp$ is a closed linear subspace of G_0.
2) $M \subset M^{\perp\perp}$
3) If $M \subset N$ then $N^\perp \subset M^\perp$.
4) $M^{\perp\perp\perp} = M^\perp$
5) $M^{\perp\perp}$ is the smallest closed linear subspace of G_0 containing M.

Exercise. If M is a subset of G_0 and [M] denotes the subspace of G_0 spanned by M then the orthogonal complement of the closure of [M] is $M^\perp$.

Exercise. If M is a subset of G_0 and $M^\perp = \{0\}$ then M is dense in G_0.

Exercise. Suppose that M is a closed linear subspace of G_0. Let P be the subset of

$G_0 \times G_0$ to which (x, y) belongs only if y is in M and $N_0(x - y) \leq N_0(x - z)$ for all z in M. Show that P is a continuous linear transformation of G_0, $P^2 = P$, and $Q_0(Px, y) = Q_0(x, Py)$ for all (x, y) in $\boldsymbol{D}(P) \times \boldsymbol{D}(P)$, where $\boldsymbol{D}(P)$ denotes the domain of P. If $\{G_0, Q_0\}$ is complete then P is said to be an <u>orthogonal projection</u> of G_0 onto M.

Suppose that each of $\{G_1, Q_1\}$ and $\{G_2, Q_2\}$ is an inner product space and A is a linear function from G_1 into G_2. The range of A will be denoted by $\boldsymbol{R}(A)$ and the null space of A is denoted by $\boldsymbol{N}(A)$. The <u>adjoint</u> of A, denoted by A*, is the subset of $G_2 \times G_1$ to which (y, z) belongs only if $Q_1(x, z) = Q_2(Ax, y)$ for each x in G_1. Here z is denoted by A*y.

<u>Exercise</u>. Suppose that each of $\{G_1, Q_1\}$ and $\{G_2, Q_2\}$ is an inner produce space, A is a linear function from G_1 into G_2, and the domain of A* is dense in $\{G_2, Q_2\}$. If x_0 is in G_1 and $N_2(Ax_0) > c > 0$, there is a positive number r such that if x is in G_1 and $N_1(x - x_0) < r$ then $N_2(Ax) > c$. Moreover, it follows that if G_1 is complete with respect to N_1 then A is continuous.

<u>Theorem</u> 1.12. If each of $\{G_1, Q_1\}$ and $\{G_2, Q_2\}$ is an inner product space and A is a continuous linear function from G_1 into G_2, then

1. $[\boldsymbol{R}(A)]^{\perp} = \mathrm{cl}(N(A^*))$
2. $\mathrm{cl}(R(A)) = [\boldsymbol{N}(A^*)]^{\perp}$
3. $[\boldsymbol{R}(A^*)]^{\perp} = \mathrm{cl}(N(A))$
4. $\mathrm{cl}(R(A^*)) = [\boldsymbol{N}(A)]^{\perp}$

<u>Theorem</u> 1.13. If each of $\{G_1, Q_1\}$ and $\{G_2, Q_2\}$ is an inner product space and A is a continuous linear function from G_1 into G_2 then A is reversible on $\boldsymbol{R}(A^*)$.

<u>Exercise</u>. If each of $\{G_1, Q_1\}$ and $\{G_2, Q_2\}$ is a complete inner product space and A is a bounded linear transformation from G_1 onto G_2 then $\boldsymbol{N}(A^*) = \{0\}$.

<u>Theorem</u> 1.14. If each of $\{G_1, Q_1\}$ and $\{G_2, Q_2\}$ is a complete inner product space

and A is a continuous, linear, reversible function from G_1 onto G_2, then A^{-1} is a continuous linear functon from G_2 onto G_1.

Suppose that $\{G_0, Q_0\}$ is a complete inner product space, E is a set, and $\{G_1, Q_1\}$ is an inner product space of functions from E into G_0. A function K from $E \times E$ into the continuous linear transformations of $\{G_0, Q_0\}$ is called a reproducing kernel for $\{G_1, Q_1\}$ provided [10]

1) Each function of the form K(, t) x, t in E and x in G_0, is in G_1.
2) If f is in G_1, t is in E, and x is in G_0 then $Q_1(f, K(, t)x) = Q_0(f(t), x)$.

Exercise. If an inner product space has a kernel, that kernel is unique.

Theorem 1.15. If K is a function from $E \times E$ into the continuous linear transformations of $\{G_0, Q_0\}$ such that, for each sequence $\{s_p, x_p\}_{p=0,n}$ in $S \times G_0$, $\Sigma_{p=0}\Sigma_{q=0} Q_0(K(s_p, s_q)x_q, x_p) \geq 0$, then there is a unique complete inner product space $\{G_1, Q_1\}$ of functions from E into G_0 which has K as a reproducing kernel, i.e., K determines a unique complete RKH space.

4. Reproducing Kernel Hilbert Space Setting for Hereditary Systems

The setting for most of the results presented in these notes is a reproducing kernel Hilbert space which we now introduce [1]. Several mathematical advantages accrue to us because of this setting. First, the operators of interest have "nice" representations, which is important for numerical solutions. Second, the computation of required adjoints is facilitated. Finally, we can impose point conditions in control problems.

The integration theory that Hellinger introduced in his 1907 dissertation [7] is largely ignored in the literature today. For most, the Lebesgue theory (or abstract integration theory) and the spaces of Sobolev [2, 3] provide the analytical setting and tools for system analysis. However, in the approach that we have adopted the technical prerequisites are smaller and the setting can be more easily tailored to the specific problem at hand.

For the remaining part of these notes we will assume that $\{X, |\cdot|\}$ is a complete inner product space, i.e., $|\cdot|$ is an inner product norm $<\cdot, \cdot>^{1/2}$. The deterministic control problems to be considered will typically require X to be finite dimensional Euclidean space, but we need to allow for X to be infinite dimensional in order to handle stochastic problems.

The space of Hellinger integrable functions [10]. Let k be a nondecreasing function from S to the numbers (real) [u, v] a subinterval of S, and f a function from S into X. If there exists a number b such that $\Sigma_{p=1,n} \mid df(s_{p-1}, s_p) \mid^2 / dk(s_{p-1}, s_p) \leq b$ for each partition $\{s_p\}_{p=0,n}$ of [u, v], then we say that f is *Hellinger integrable on [u, v] with respect to* k. We assume in the sum that if $dk(s_{p-1}, s_p) = 0$ then $df(s_{p-1}, s_p) = 0$ and we define $\mid df(s_{p-1}, s_p) \mid^2 / dk(s_{p-1}, s_p)$ to be 0. The least number b for which the above inequality holds is denoted by $\int_{[u,v]} \mid df \mid^2 / dk$.

The space G_H of Hellinger integrable functions with respect to k consists of all functions f from S into X such that $\int_{[u,v]} \mid df \mid^2 / dk$ exists for each subinterval [u, v] of S. The least upper bound of the integrals $\int_{[u,v]} \mid df \mid^2 / dk$ for [u, v] contained in S, if it exists, is denoted by $\int_S \mid df \mid^2 / dk$. If S is infinite, we define G_∞ to be the set of all Hellinger integrable functions f with respect to k such that $\int_S \mid df \mid^2 / dk$ is finite.

The following result is fundamental to the development of a reproducing kernel Hilbert (RKH) space of Hellinger integrable functions. It also plays a central role in obtaining finite sum approximations of Hellinger integrals.

Lemma 1.16. If {u, v, w} is an increasing sequence in S then $\mid df(u, w) \mid^2 / dk(u, w) \leq \mid df(u, v) \mid^2 / dk(u, v) + \mid df(v, w) \mid^2 / dk(v, w)$. That is, approximating sums increase under refinement.

Proof. The lemma is an immediate consequence of the Cauchy-Schwarz inequality

$$
\begin{aligned}
\mid df(u, w) \mid^2 / dk(u, w) &\leq \{ \mid df(u, v) \mid (dk(u, v)/dk(u, v))^{1/2} \\
&\quad + \mid df(v, w) \mid (dk(v, w)/ dk(v, w))^{1/2} \}^2 / dk(u, w) \\
&\leq \mid df(u, v) \mid^2 / dk(u, v) + \mid df(v, w) + \mid df(v, w) \mid^2 / dk(u, v) \qquad \square
\end{aligned}
$$

Exercise. Suppose that k is the identity function. If f is in G_H then f is absolutely continuous.

Our immediate goal is to obtain an inner product on G_H. Thinking of $\int_S |df|^2/dk$ as the obvious candidate for a norm, it is natural to investigate integrals of the form $\int_S \langle df, dg\rangle/dk$. We now show that such integrals exist for f, g in G_H, and may be computed through refinement of partitions.

Theorem 1.17. If each of f and g is in G_H and [u, v] is a subinterval of S then there is a number J with the property that for each positive number ε there is a partition $\{s_p\}$ of [u, v] such that

$$| J - \sum_{q=1}^{m} \langle df(t_{q-1}, t_q), dg(t_{q-1}, t_q)\rangle /dk(t_{q-1}, t_q) | < \varepsilon$$

for each refinement $\{t_q\}_{q=0,m}$ of $\{s_p\}$. Here J will be denoted as $\int_{[u,v]} \langle df, dg\rangle/dk$.

Proof. If h is in G_H then $\int_{[u,v]} |dh|^2/dk$ is the least number b such that $\Sigma_{p=1,n} | dh(s_{p-1}, s_p) |^2/dk(s_{p-1}, s_p) \le b$ for all partitions $\{s_p\}_{p=0,n}$ of [u, v]. The lemma shows that $\int_{[u,v]} |dh|^2/dk$ is a limit through refinements, i.e., a Hellinger integral. Notice that the set G_H of Hellinger integrable functions is a linear space. If each of α and ß is a complex number and $\{s_p\}_{p=0,n}$ is a partition of [u, v] then

$$\sum_{p=1}^{n} | \alpha f(s_p) + \beta g(s_p) - \alpha f(s_{p-1}) - \beta g(s_{p-1}) |^2 /dk(s_{p-1}, s_p)$$

$$\le 2(\alpha^2 + \beta^2)(\sum_{p=1}^{n} (| df (s_{p-1}, s_p) |^2 + | dg(s_{p-1}, s_p) |^2) / dk(s_{p-1}, s_p))$$

$$\le 2(\alpha^2 + \beta^2)(\int_u^v | df |^2 / dk + \int_u^v | dg |^2 / dk)$$

For q = 1, 2, 3, 4, let

$$J_q = \int_u^v | d(f + i^q g) |^2 / dk$$

$$J = \sum_{q=1}^{4} (i^q/4) J_q$$

where i is the complex number. If ε is a positive number there is a partition $\{s_p\}_{p=0,n}$ such that if $\{t_j\}_{j=0,m}$ refines s then

$$| J_q - \sum_{j=1}^{m} | f(t_j) + i^q g(t_j) - f(t_{j-1}) - i^q g(t_{j-1}) |^2 /dk(t_{j-1}) | \leq \varepsilon$$

for q = 1, 2, 3, 4. Hence

$$| J - \sum_{j=1}^{m} <df(t_{j-1}, t_j), dg(t_{j-1}, t_j)> /dk(t_{j-1}, t_j) | \leq \sum_{q=1}^{4} (1/4) | J_q - \sum_{j=1}^{m} | f(t_j)$$

$$+ i^q g(t_j) - f(t_{j-1}) - i^q g(t_{j-1}) |^2 /dk(t_{j-1}, t_j) < \varepsilon \quad \square$$

If each of f and g is in G_H and both $\int_S | df |^2/dk$ and $\int_S | dg |^2/dk$ exist, let $Q_H(f, g) = <f(-r), g(-r)> + \int_S <df, dg>/dk$, where

$$\int_S <df, dg> /dk = \begin{cases} \int_{-r}^{T} <df, dg> /dk & \text{if } S = [-r, T] \\ \lim_{T\to\infty} \int_{-r}^{T} <df, dg> /dk & \text{if } S = [-r, \infty] \end{cases}$$

As usual, $N_H(f) = Q_H(f, f)^{1/2}$.

Theorem 1.18. For each T in S, $\{P_T G_H, Q_H\}$ is a complete inner product space.

Proof. Note that if f is in $P_T G_H$ then

$$| f(s) | \leq | f(-r) | + | df(-r, s) | (dk(-r, s)/dk(-r, s))^{1/2}$$
$$\leq \{1 + dk(-r, s)\}^{1/2}\{ | f(-r) |^2 + | df(-r, s) |^2/dk(-r, s)\}^{1/2}$$
$$\leq \{1 + dk(-r, s)\}^{1/2} N_H(P_T f)$$

for each s in [-r, T]. Therefore if $\{f_p\}_{p=0,\infty}$ is a Cauchy sequence in $P_T G_H$ there is a function g from S into X such that g(s) is the limit of $\{f_p(s)\}_{p=0,\infty}$ for each s in S. Furthermore, if $\varepsilon > 0$ and $\{s_p\}_{p=0,n}$ is an increasing sequence in [-r, T] then there is a positive integer m such that $N_H(P_T(f_m - f_{m+q}))^2 < \varepsilon/4$ for each nonnegative integer q. Also there is a nonnegative integer q such that

$$\sum_{p=1}^{n} | g(s_p) - f_{m+q}(s_p) - g(s_{p-1}) + f_{m+q}(s_{p-1}) |^2 /dk(s_{p-1}, s_p) < \varepsilon/4$$

Hence

$$\sum_{p=1}^{n} | g(s_p) - f_m(s_p) - g(s_{p-1}) + f_m(s_{p-1}) |^2 / dk(s_{p-1}, s_p)$$

$$\leq 2 \sum_{p=1}^{n} | g(s_p) - f_{m+q}(s_p) - g(s_{p-1}) + f_{m+q}(s_{p-1}) |^2 / dk(s_{p-1}, s_p)$$

$$+ 2N_H^2(f_m - f_{m+q}) \leq \varepsilon$$

i.e., $g - f_m$ is in $P_T G_H$ or g is in $P_T G_H$ and is the limit of $\{f_p\}_{p=0,\infty}$ with respect to $N_H(P_T\cdot)$. □

Exercise. Show that $| f(t) | \leq (1 + dk(-r, t))^{1/2} N_H(P_T f)$

Exercise. In the case that $S = [-r, \infty]$ show that $\{G_\infty, Q_H\}$ is a complete inner product space.

Exercise. Suppose $r = 0$ and G_0 is all members of G_H satisfying $f(0) = 0$. Show that $\{G_0, Q_H\}$ may be characterized as the space of absolutely continuous functions on $[0, \infty]$ with inner product $\langle f, g \rangle = \int_{[0,\infty]} f'(s)\, g'(s)\, ds$. This example is well known as an RKH space representation of Brownian motion (see [4] or Chapter IV, Section 3 of these notes).

Exercise. If f belongs to G_H and u, v are in S, show that $| f(v) - f(u) |^2 \leq (k(v) - k(u)) \int_{[u,v]} | df |^2 / dk$. Conclude that f is absolutely continuous with respect to k.

We have now shown that $\{P_T G_H, Q_H\}$ is a Hilbert space for each T in S. It is, in fact, an RKH space with a particularly nice reproducing kernel. Before revealing the kernel, we pause to relate $\{P_T G_H, Q_H\}$ to more standard function spaces.

As shown in the preceding exercise, $P_T G_H$ is a subspace of the absolutely continuous functions with respect to k on $S = [-r, T]$, which is in turn a subspace of the functions of bounded variation on S. This is, in turn, a subspace of the quasi-continuous functions on S. Letting $| f |_\infty$ and var(f) denote the supremum and variational norms of f on S, respesctively, it follows that $|f|_\infty \le \text{var}(f) \le (1 + k(T) - k(-r))^{1/2} N_H(f)$. In addition, G_H is dense, with respect to the variational norm, in the space of functions absolutely continuous with respect to k.

We now return to the identification of $\{P_T G_H, Q_H\}$ as a RKH space. Let K denote the function from $S \times S$ into the continuous linear transformations of X given by

$$K(u, v) = \begin{cases} (k(u) - k(-r) + 1) \cdot I & -r \le u \le v \le T \\ (k(v) - k(-r) + 1) \cdot I & -r \le v \le u \le T \end{cases}$$

Theorem 1.18. For each T in S, K is a reproducing kernel for $\{P_T G_H, Q_H\}$, i.e., $K(\cdot, u)x$ is in $P_T G_H$, for each u in [-r, T] and x in X, and $Q_H(f, K(\cdot, u)x) = \langle f(u), x\rangle$ for each f in $P_T G_H$, u in [-r, T], and x in X.

Proof. If $\{s_p\}_{p=0,n}$ is an increasing sequence in S and $s_q = u$ for some q = 0, 1, $\cdots$, n then

$$\sum_{p=1}^{n} |K(s_p, u)x - K(s_{p-1}, u)x|^2 / dk(s_{p-1}, s_p) = \sum_{p=1}^{q} dk(s_{p-1}, s_p)\,|x|^2 \le dk(-r, u)\,|x|^2$$

Therefore $K(\cdot, u)x$ is in G_H. Also for f in $P_T G_H$, we have

$$\sum_{p=1}^{q} \langle df(s_{p-1}, s_p), K(s_p, u)x - K(s_{p-1}, u)x\rangle / dk(s_{p-1}, s_p)$$

$$= \sum_{p=1}^{q} \langle df(s_{p-1}, s_p), x\rangle = \langle df(s_0, u), x\rangle$$

It follows that $Q_H(f, K(, u)x) = \langle f(u), x\rangle$. □

Exercise. If $S = [-r, \infty)$ then $\{G_\infty, Q_H\}$ is a reproducing kernel Hilbert space with kernel K.

Theorem 1.20. If T is in S^+ and A is a continuous linear transformation of $\{P_T G_H, Q_H\}$ then there is a function L from $S \times S$ into the continuous linear transformations of X such that $<[Af](u), x> = Q_H(f, L(\cdot, u)x)$, for each f in $P_T G_H$, u in S, and x in X. L is called the <u>matrix representation</u> of A on $P_T G_H$.

Here $<[Af](u), x> = Q_H(Af, K(\cdot, u)x) = Q_H(f, A^*K(\cdot, u)x)$ and hence $L(v, u)x = [A^*K(\cdot, u)x](v)$ for each u, v in [-r, T] and x in X. Note that K agrees with the matrix representation of I on $P_T G_H$ on $[-r, T] \times [-r, T]$.

<u>Exercise</u>. Given that L is the matrix representation of A on $P_T G_H$, show that the matrix representation L_1 of A^* is given by $L_1(u, v) = L^*(v, u)$.

<u>Representation of elements of *A* and *B*</u>. We are now in a position to offer a different answer to the question "How big is the class of hereditary systems?" The idea is to characterize those matrix representations of continuous linear operators on G_H which correspond to restrictions of elements of ***A*** and ***B*** to G_H. Further, we want to see how various system properties, for instance time invariance, can be related to properties of the matrix representation.

The linear space of functions G_H is associated with a particular nondecreasing function k. In most of what follows we will be concentrating on a single system at any time. This means that we can pick out one nondecreasing function which will serve for all of the members of ***A*** and ***B*** which enter that particular discussion. Thus there will be only one inner product space of the type G_H in any discussion. Finally, we require that $| f(v) - f(u) | \leq N_v(P_v f - P_u f)$ for $u \leq v$ in S^+ and f in G. Also, $| f(u) | \leq N_u(f) \leq (1 + k(u) - k(-r)^{1/2} N_H(f)$ for u in S and f in G_H. Note that our standard examples satisfy these conditions.

<u>Theorem</u> 1.21. If B is in ***B***, k is a variation for B and G_H is the space of functions from S into X which are Hellinger integrable with respect to k, then B maps G into G_H. Furthermore, if G_H is contained in G and $A = (I - B)^{-1}$ then A maps G_H onto G_H.

The requirement that G_H be contained in G is a little delicate, i.e., it depends on the choice of k. Again, for the standard examples there will be a "natural" choice for k which yields this condition. We will <u>not</u> impose this condition in general, but we will require that it hold for the remainder of this chapter.

<u>Proof</u>. If f is in G, [a, b] is subinterval of S, and $\varepsilon > 0$ then there is a parti-

tion $\{s_p\}_{p=0,n}$ of [a, b] such that

$$\sum_{p=1}^{n} | [Bf](s_p) - [Bf](s_{p-1}) |^2 /dk(s_{p-1}, s_p)$$

$$\leq \sum_{p=1}^{n} N_{s(p)}(f)^2 \, dk(s_{p-1}, s_p)$$

$$\leq (R) \int_{[a,b]} N_x(f)^2 \, dk(x) + \varepsilon$$

Thus, for each increasing sequence $\{s_p\}_{p=0,n}$ in S,

$$\sum_{p=1}^{n} | [Bf](s_p) - [Bf](s_{p-1}) |^2 /dk(s_{p-1}, s_p) \leq (R) \int_{[s(0),s(n)]} N_x(f)^2 \, dk(x)$$

i.e. Bf is in G_H. If (f, g) is in $G \times G$ and $h = f + Bh = (I - B)^{-1}f$ then h is in G_H if and only if f is also. □

Theorem 1.22. If D is in ***A*** or ***B*** and T is in S then the restriction of P_TD to P_TG_H is a continuous linear transformation of $\{P_TG_H, Q_H\}$.

Proof. The inequalities in the proof of Theorem 1.21 show that $N_H(P_TBf)^2 \leq$ (R) $\int_{[-r,T]} N_x(f)^2 \, dk(x) \leq (1 + k(T) - k(-r)) \, dk(-r, T) N_H(P_Tf)^2$, for each B in ***B*** and f in P_TG_H. □

Exercise. If D is in A or B, $T < T'$ in S, L is the matrix representation of P_TD on $\{P_TG_H, Q_H\}$, and L′ is the matrix representation of $P_{T'}D$ on $\{P_{T'}G_H, Q_H\}$ then $L'(u, u) = L(u, v)$ for $-r \leq u, \; v \leq T$. Thus we speak of the matrix representation of D without reference to T.

Exercise. Suppose that S = [0, T], where T is a positive number, $k = I + 1$, and B is the continuous linear transformation of $\{G_H, Q_H\}$ defined by $[Bf](u) = \int_{[0,u]} f \, dI$, for each f in G_H and u is S. Find a formula for B*.

Exercise. If each of D_1 and D_2 is in $\boldsymbol{A} \cup \boldsymbol{B}$, and L_i is the matrix representation of D_i, for i = 1, 2, then

1. $L_1 + L_2$ is the matrix representation of $D_1 + D_2$

2. When $S = [0, T]$, the matrix representation L_{12} of D_1D_2 is given by $L_{12}(u, v) = L_2(u, 0)L_1(0, v) + \int_{[0,T]} dL_2(u, \cdot)\, dL_1(\cdot, v)/dk$, for each u, v in S
3. Assume $S = [0, T]$, D_1 is in $\boldsymbol{A}$, $D_2 = D_1^{-1}$, $k(t) = t + 1$ for each t in S, and both $\partial L_1(t, v)/\partial t$ and $\partial^2 L_1(t, v)/\partial t^2$ are continuous on S for each v in S. Show that for $u \le v$ in S, $L_2(u, v) = \partial L_1(t, v)/\partial t \mid_{t=v} - \partial L_1(t, v)/\partial t \mid_{t=0} - \int_{[u,v]} L_2(u, t)(\partial^2 L_1(t, v)/\partial t^2)\, dt$, i.e., $L_2(u, \cdot)$ satisfies an integral equation given in terms of L_1.

For the rest of the subsection suppose that G is in the space of functions from S to X of bounded variation, $N_u(f) = | f(-r) | + \int_{[-r,u]} | df |$ for each f in G and u in S, and

$$[Pf_u](x) = \begin{cases} f(x) & x \le u \\ f(u) & u \le x \end{cases}$$

Theorem 1.23. Assume $S = [-r, T]$. A function L from $S \times S$ into the continuous linear transformations of X is the matrix representation of the restriction of some member of $\boldsymbol{B}$ to G_H only in case

1. $L(v, u) = 0$ for v in S and $-r \le u \le 0$
2. $L(v, u) = L(u, u)$ for $0 \le u \le v$
3. There is an ordered pair (k_1, k_2) of nondecreasing functions from S to the numbers such that each of k_1 and k_2 is Hellinger integrable with respect to k and

$$| L(b, v)x - L(a, u)x - L(b, u)x + L(a, u)x | \le | x |\, dk_1(a, b)\, dk_2(u, v)$$

for each x in X and ordered pair ([a, b], [u, v]) of subintervals of S.

Proof. Suppose that L is the matrix representation of the restriction of some element B of $\boldsymbol{B}$ to G_H. Then

$$\begin{aligned} \langle y, L(v, u)x\rangle &= Q_H(K(\ , v)y, L(\ , u)x) \\ &= \langle [BK(\ , v)y](u), x\rangle \\ &= 0 \end{aligned}$$

when $u \leq 0$ for each v in S and (x, y) in $X \times X$, i.e., $L(v, u) = 0$. Furthermore, if $0 < u \leq v$ then

$$\begin{aligned} <y, L(v, u)x> &= <[BP_u K(\ , v)y](u), x> \\ &= <[BK(\ , u)y](u), x> \\ &= <y, L(u, u)x> \end{aligned}$$

i.e., $L(v, u) = L(u, u)$. Finally, there is a number c such that

$$\begin{aligned} &| <y, L(b, v)x - L(a, v)x - L(b, u)x + L(a, u)x> | \\ &\qquad \leq c \, | x | \int_u^v N_t(K(\ , b)y - K(\ , a)y) \; dk(t) \\ &\qquad \leq c \, | x | \, | y | \, dk(a, b) \, dk(u, v) \end{aligned}$$

Hence L satisfies conditions (1) - (3).

Suppose that L is a function from $S \times S$ into the continuous linear transformations of X satisfying conditions (1) - (3). Note that $L(\cdot, u)x$ is in G_H for each u in S and x in X. Let B denote the linear transformation of G_H defined by $<[Bf](u), x> = Q_H(f, L(\cdot, u)x)$, for each f in G_H, x in X, and u in S. Clearly, $[Bf](u) = 0$ for each f in G_H if $u \leq 0$. If (k_1, k_2) is an ordered pair of nondecreasing functions from S to the numbers satisfying (3) then there is a number c such that $dk_1(u, v)$ and $dk_2(u, v)$ are bounded above by $c \, dk(u, v)$ for each subinterval [u, v] of S. If $\{s_p\}_{p=0,n}$ is a partition of [u, v] then

$$\begin{aligned} &| <[Bf](v) - [Bf](u), x> | \\ &\qquad \leq \sum_{p=1}^{n} | \int_{[-r,s(p)]} <df, d(L(\cdot, s_p)x - L(\cdot, s_{p-1})x) > /dk | \\ &\qquad \leq c^2 \, | x | \sum_{p=1}^{n} N_{s(p)}(f) \; dk(s_{p-1}, s_p) \\ &\qquad \leq c^2 \, | x | \int_u^v N_t(f) \; dk(t) \end{aligned}$$

In addition, B can be extended to a continuous linear transformation of G contained in ***B***. □

Exercise. Assume that $S = [-r, T]$ and $k = I + r + 1$. If B is in ***B*** and $N_w(P_vBf - P_uBf) \le \int_{[u,v]} |f|\, dI$, for each f in G and $u \le v \le w$ in S then the transformation E of G_H defined by $[Ef](u) = B^*f - K(\cdot, T)[B^*f](0)$ for each f in G_H and u in S can be extended to an element of ***B***.

System properties and related properties of matrix representations. Our purpose in this subsection is to relate various elementary system properties to the matrix representations of the system operators. We assume for the rest of the chapter that $S = [-r, \infty]$ and

$$k(t) = \begin{cases} t + r + 1 & -r \le t < 0 \\ t + 2r + 1 & 0 \le t \end{cases}$$

The size of the jump at 0 is irrelevant, but making it r allows for the possibility that $r = 0$.

For each positive number b and f in G let

$$[S_bf](t) = \begin{cases} 0 & -r \le t < b \\ f(t - b) & b \le t \end{cases}$$

and

$$[S_{-b}f](t) = \begin{cases} 0 & -r \le t < 0 \\ f(t + b) & 0 \le t \end{cases}$$

The operator S_b 'shifts' the graph of a function in $(I - P_0)G$ to the right and S_{-b} 'shifts' the graph of a function in $(I - P_b)G$ to the left. Consequently, we say that the operator A is time invariant provided $S_{-b}AS_bf = Af$ for each f in $(I - P_0)G$ and $b > 0$. To illustrate the concept, let A denote the set of ordered pairs (f, h) in $G \times G$ such that

$$h(t) = \begin{cases} f(t) & -r \le t \le 0 \\ f(t) + \alpha\int_0^t h(u)\, du + \beta\int_0^t h(u - r)\, du & 0 < t \end{cases}$$

where α and ß are d × d matrices. If f is in $(I - P_0)G$ and $b > 0$ then

$$[AS_b f](t) = \begin{cases} 0 & -r \le t \le b \\ f(t-b) + \alpha\int_b^t [AS_b f](u)\,du + \beta\int_b^t [AS_b f](u - r)\,du & b \le t \end{cases}$$

and

$$[S_{-b}AS_b f](t) = \begin{cases} 0 & -r \le t \le 0 \\ f(t) + \alpha\int_0^{t+b} [AS_b f](u)\,du + \beta\int_0^{t+b} [AS_b f]\,(u - r)\,du & 0 \le t \end{cases}$$

$$= \begin{cases} 0 & -r \le t \le 0 \\ f(t) + \alpha\int_0^{t} [S_{-b}AS_b f](u)\,du + \beta\int_0^{t} [S_{-b}AS_b f]\,(u - r)\,du & 0 \le t \end{cases}$$

Hence $S_{-b}AS_b f = Af$, i.e., A is time invariant.

Theorem 1.24. A linear transformation A of G_H with matrix representation L is time invariant if and only if $L(v, w) - L(u, w) = L(v + b, w + b) - L(u + b, w + b)$ for each $0 \le u \le v \le w$ and positive number b.

Proof. Note that if u and v are nonnegative numbers, $K(t - b, v) - K(t - b, u) = K(t, v + b) - K(t, u + b)$. Hence, if A is time invariant,

$$\begin{aligned} \langle y, L(v, w)x - L(u, w)x\rangle &= \langle [A(K(\cdot, v)y - K(\cdot, u)y)](w), x\rangle \\ &= \langle [A(K(\cdot, v + b)y - K(, u + b)y)](w + b), x\rangle \\ &= \langle y, L(v + b, w + b)x - L(u + b, w + b)x\rangle \end{aligned}$$

i.e., $L(v, w) - L(u, w) = L(v + b, w + b) - L(u + b, w + b)$. The other implication follows upon rearranging the last set of equations. □

Assuming that A is in $\mathcal{A}$, the matrix representation L of A contains all of the information of the qualitative behavior of the system response $h = Af$. If $d = 1$, i.e., $X = R$, $S = [0, \infty]$, and the system is given formally by $h(t) = f(t) + \alpha\int_{[0,t]} h(s)\,ds$, where α is a scalar, then $[Af](t) = e^{\alpha t} f(0) + \int_{[0,t]} e^{\alpha(t-s)}\,df(s)$. For this invariant system we can use the variation of constants representation of A to draw some conclusions about the qualitative behavior of the system. Since

$$|\,[Af](t)\,| \le e^{\alpha t}\,|\,f(0)\,| + \left(\int_0^t e^{2\alpha(t-s)}\,ds\right)^{1/2} \left(\int_0^t |\,df\,|^2/dk\right)^{1/2}$$

for each f in G_H and t in S, [Af](S) is bounded if $\alpha < 0$ and $\int_{[0,\infty]} |\,df\,|^2/dk$ exists, i.e., if finite. How do we get the same information from the matrix representation L of A?

If $s \le t$ then

$$L(s, t) = e^{\alpha t}K(0, s) + \int_0^t e^{\alpha(t-x)}\,dK(x, s)$$

$$= e^{\alpha t} + \int_0^s e^{\alpha(t-x)}\,dx$$

$$= e^{\alpha t} + -(1/\alpha)(e^{\alpha(t-s)} - e^{\alpha t})$$

Hence $\partial L(s, t)/\partial s\,|_{s=0} = e^{\alpha(t-s)}\,|_{s=0} = e^{\alpha t}$, i.e.,

$$[Af](t) = f(0)L(0, t) + \int_0^t df(s)dL(s, t)/ds$$

$$= f(0)L(0, t) + \int_0^t (\partial L(s, t)/\partial s)\,df(s)$$

$$= f(0)L(0, t) + \int_0^t \partial L(u, v)/\partial u\,|_{(0,t-s)}\,df(s)$$

Clearly, if L(0, s) is bounded and $\int_{[0,\infty]} |\,\partial L(u, v)/\partial u\,|_{(0,\,s)}\,|^2$ is bounded then [Af](s) is bounded if $\int_{[0,\infty]} |\,df\,|^2/dk$ is finite.

The crucial point is that for time invariant systems we want to draw conclusions about the system from properties of $L(0, \cdot)$ and $\partial L(u, v)/\partial u]\,|_{(0,\,\cdot)}$. The following theorem is an immediate consequence of the Mean Value Theorem and Theorem 1.23.

<u>Theorem</u> 1.25. If X = R, S = [0, ∞), A is time invariant, and $\partial L(u, v)/\partial u]\,|_{(0,\,\cdot)}$ is continuous on S then

$$[Af](t) = f(0)L(0, t) + \int_0^t \partial L(u, v)/\partial u \,|_{(0,t-s)}\, df(s)$$

for each f in G_H and t in S.

<u>Corollary</u> 1.26. If L(0, S) is bounded and $\int_{[0,\infty)} |\, \partial L(u, v)/\partial u \,|_{(0, s)}\,|^2\, ds$ is finite then [Af](S) is bounded for each f in G_H with $\int_{[0,\infty)} |\, df\,|^2/dk$ finite.

<u>Exercise</u>. If $X = R^d$, $d > 1$, $S = [0, \infty)$, A is time invariant, and $\partial L(u, v)/\partial u \,|_{(0,\cdot)} x$ is continuous on S for each x in X then

$$\langle [Af](t), x\rangle = \langle f(0), L(0, t)x\rangle + \int_0^t \langle df(s), \ \partial L(u, v)/\partial u \,|_{(0,t-s)} x\rangle$$

for each f in G_H, t in S, and x in X.

<u>Exercise</u>. If L(0, S)x is bounded and $\int_{[0,\infty)} |\, \partial L(u, v)/\partial u \,|_{(0, s)}\, x|^2\, ds$ is finite for each x in X then [Af](S) is bounded for each f in G_H with $\int_{[0,\infty)} |\, df\,|^2/dk$ finite.

Summary

We have set up a framework for the analysis of structured hereditary systems. This framework consists of two classes of operators with which to model system dynamics, and a class of function spaces in which to solve problems. We have illustrated some of the basic ideas of our method, deriving several properties of the operator spaces ***A*** and ***B*** and the spaces of Hellinger integrable functions. Along with providing the support for our subsequent analysis, these results should indicate the advantages of our method.

The motivation for the defining inequalities for the classes ***A/B*** comes from the theory of integral equations. We believe there are several advantages to the ***A/B*** system description when compared to alternative approaches such as semigroups or abstract evolution equation descriptions.

First, the ***A/B*** description is closer to traditional ordinary differential equation, integral equation, or functional differential equation descriptions in the sense that less abstract mathematical machinery is needed to establish basic system properties. For instance, the existence and uniqueness arguments given in this chapter parallel the usual arguments.

Second, the algebraic properties of the ***A*/*B*** operators are useful for the analysis of structured systems. In terms of signal diagrams we can think of putting components together to form more complex systems or of decomposing a system into simpler components. This may be the most significant aspect of our work.

Third, the ***A*/*B*** description makes it convenient to formulate and analyze system problems in a reproducing kernel Hilbert space setting. As indicated in the chapter, system dynamics may be finite or infinite dimensional, no distinction has thus far been made between random and deterministic inputs, and the system may be run on finite or infinite time intervals. Consequently, in our approach modeling and analysis occur at the same level of abstraction.

We have not yet seen the full power of a reproducing kernel Hilbert space. In later chapters, the mathematical elegance of the RKH space setting will be evident, principally through the formulation of point conditions in control problems and the straightforward calculation of explicit formulas for adjoint and feedback operators.

Several advantages of the space of Hellinger integrable functions are already apparent. From a modeling viewpoint, we have a great deal of flexibility. As noted above, the ***A*/*B*** description of hereditary systems is quite general, and we have shown that operators from ***B*** map G into G_H, and operators from *A* map G_H onto G_H. Within our setting, the freedom to choose k in a particular application provides additional flexibility for the analysis of system problems. For example, if k is the identity, G_H is the space of absolutely continuous functions on S with square (Lebesque) integrable derivative, and N_H is equivalent to the Sobolev norm. Our analysis then proceeds in $W_{1,2}$ without the introduction of measure theoretic concepts.

The Hellinger integral gives us several computational benefits as well. The primary advantage is the existence of readily computable matrix representations of operators in ***A*/*B***. Further, the Hellinger norm N_H is approximated from below by finite sums, with convergence through refinement of partitions. Some numerical applications of RKH space theory are gathered in [18].

The general nature of the theory given in this chapter offers great hope for developments beyond what is detailed here. The specific applications to be pursued in later chapters are outlined as follows.

In Chapter II, we will see that system concepts such as inputs, outputs, feedback, and observations have easily obtained ***A*/*B*** descriptions. In Chapter III, projection methods are used to provide finite dimensional approximations to system operators and to analyze identification/parameter estimation problems and the numerical solution to optimal control problems. In Chapter IV, we will see that many problems for stochastic systems can be analyzed within the same framework as developed in the preceeding chapters. For systems with state dependent noise there is a natural extension of our ***A/B*** description. In Chapter V, the ***A/B*** description via signal diagrams is used to describe interconnections between system components and to develop control coordination schemes for large scale systems.

Our thoughts about systems theory in general and hereditary systems in particular have been influcenced by many individuals. We have included only basic references in this chapter.

References

1. N. Aronszajn, Theory of reproducing kernels, Trans. AMS, 68(1950).

2. A. B. Balakrishnan, Applied Functional Analysis, Springer-Verlag, Berlin, 1976.

3. R. F. Curtain and A. J. Pritchard, Infinite Dimensional Linear Systems Theory, Springer-Verlag, Berlin, 1978.

4. A Feintuck and F. Saeks, Systems Theory: A Hilbert Space Approach, Academic Press, New York, 1982.

5. U. Grenander, Abstract Inference, Wiley-Interscience, New York, 1981.

6. J. K. Hale, Theory of Functional Differential Equations, Springer-Verlag, Berlin, 1977.

7. E. D. Hellinger, Die Orthogonal invarianten Quadratischer Formen von Unendlichvielen Variablen, Dissertation, Gottingen, 1907.

8. R. E. Kalman, P. L. Falb, and M. A. Arbib, Topics in Mathematical Systems Theory, McGraw-Hill, New York, 1969.

9. D. G. Luenberger, Optimization by Vector Space Methods, John Wiley, New York, 1969.

10. J. S. Mac Nerney, Hellinger integrals in inner product spaces, J. Elisha Mitchell Scientific Society, 76(1960), 252-273.

11. E. J. McShane, Stochastic Calculus and Stochastic Models, Academic Press, New York, 1974.

12. A. N. Michel and R. K. Miller, Qualitative Analysis of Large Scale Dynamical Systems, Academic Press, New York, 1977.

13. E. Parzen, An approach to time series analysis, Am. Math. Stat. 32(1961), 951-989.

14. F. S. Roberts, Discrete Mathematical Models, Prentice Hall, Englewood Cliffs, NJ, 1976.

15. D. L. Russell, Mathematics of Finite Dimensional Control Systems: Theory and Design, Marcel Dekker, New York, 1979.

16. D. D. Siljak, Large Scale Dynamic Systems: Stability and Structure, North Holland, New York, 1978.

17. Special issue on recent trends in system theory, Proceedings of the IEEE, 64(1976), 1-192.

18. H. L. Weinert, ed., Reproducing Kernel Hilbert Spaces, Hutchinson Ross Publishing Company, 1982.

19. J. C. Willems, The Analysis of Feedback Systems, The M.I.T. Press, Cambridge, Mass., 1971.

20. W. M. Wonham, Linear Multivariable Control: A Geometric Approach, Springer-Verlag, Berlin, 1974.

II
Optimal Control

1. Introduction

We are concerned in this chapter with centralized control for hereditary systems. We take up a variety of optimization problems, particularly quadratic cost problems on finite intervals with terminal constraints, using variational methods in function spaces. The reproducing kernel Hilbert spaces of Chapter I are exploited to give 'explicit' solutions for many of these problems.

A description of overall system architecture is completed in this section with the introduction of input and output processes. The system descriptions of Chapter I will then be seen as descriptions of the internal part or the dynamics of a system. The realism of system control problems is increased significantly by posing them for systems from this enriched class.

All of the optimal control problems are posed as constrained optimization problems on function spaces. These problems are discussed abstractly in Section 2 which includes a quick review of the required optimization background. In Section 3, the basic optimal control problems for hereditary systems are discussed for systems with degenerate input/output processes. The finite interval methods are then applied to the problem of system stabilization, an infinite interval problem.

We elaborate on the basic problems in the last section, adding input and observation processes. A 'checkable' controllability criterion for hereditary systems is given in terms of the matrix representation of the input/response operator. Finally, an observation process is added, and we examine the problem of achieving control using observations in place of system responses.

System Problems. On an intuitive level we will regard a hereditary system as a structure into which we put something (matter, energy, or information) at various times and which in turn puts out something at various times [23]. (A simple mental picture of some signal processor, say a radio receiver, can serve as a guide to our elaboration of this intuitive understanding.) Our structure subdivides into at least three parts. The first is the collection of inputs and a mechanism for internalizing the inputs. As a starting place, we associate with our structure a time set S, and the inputs are then realized as a collection Ω of functions from S into a set U, which we assume from the beginning is a linear space, with the property that $f(t) = 0$ for each f in Ω and $t \leq 0$ in S. The inputs are that part of the system which we are able to act on directly. The assumption that an input is zero before $t = 0$ means that we cannot influence the system before $t = 0$.

The second or internal part of the system is realized by a space G (of the type discussed previously) of functions from S into a linear space X. The system dynamics are described by an element from ***A*** or ***B***. The mechanism for internalizing an input is a causal (nonanticipatory) function ρ from Ω into G. We might as well assume that Ω is contained in $(I-P_0)G$. Usually, we can think of U as a proper subspace of X. Thus the dynamical equation becomes, for instance, of the form $h = f + Bh + \rho(u)$, where u is in Ω and $P_t\rho = P_t\rho P_t$ for each t in S. The function u is called the control and h is called the system response.

Finally, the third part of the system is a collection of outputs and a mechanism for externalizing the system responses as outputs. The outputs are realized as a collection Γ of functions from S into a set Y and the mechanism is a causal function η from G into Γ. Again, we will assume that Γ is a subspace of G and note that Y is usually a proper subspace of X. We will speak of the outputs as observations.

A central theme of these notes is that one can profit by thinking of systems as composed of more than the three parts outlined here. However, this simple picture seems best as a starting place for considering structured system problems. In this chapter we undertake only problems of centralized control, i.e., only one 'controller', and leave to Chapter V problems with more than one controller.

The controller, perhaps no more than a piece of hardware or a computer program, is to produce a system input which will force the system response to meet some objective. That the response meets the objective might have to be deduced from system observations. Frequently, from among all of the feasible controls an optimal control is sought.

System objectives are realized as conditions on the responses h, elements of G. For S an infinite interval the conditions might be no more than h be bounded or that h have a finite limit at ∞. For S a compact interval $[-r, T]$ the conditions might include a terminal condition, for instance, $h(T) = 0$ or $h(t) = 0,\ T - r \le t \le T$.

Complications can arise from the input and observation processes. The controls and observations might be restricted or partial, i.e., $U \neq X$ and $Y \neq X$, or There may be delays in the input function ρ or output function η. In a simple case, for instance, the controller might have to deduce $h(T)$ based on observations $h(t)$, $0 \le t \le T - r$. Clearly, we need to study the question of when system objectives can be met using information about the responses gained from the observations.

Costs or benefits are implicit in the idea of optimization. We want to choose from among the feasible controls, those which yield system objectives, one which minimizes costs or maximizes benefits. Optimal control problems are usually posed in terms of minimizing costs. The costs may be real or introduced artificially to meet some 'external' design criterion.

We will formulate each optimal control problem as a constrained optimization problem, the elements of which are a linear space of functions, a set of constraints, and a cost functional. The background required for solving such problems will be reviewed in the next section. For now, we will be content with setting up a sample problem to illustrate the balance that obtains between these three elements.

Consider the formal system

$$h(t) = f(t) + \alpha \int_0^t h(s)\, ds + \beta u(t)$$

$$y(t) = \gamma h(t),$$

where α is $d \times d$ matrix, β is $d \times n$, and γ is $m \times d$. Recall that we consider the system formal until the function spaces are specified. Necessarily, U is the space of n-tuples, X the space of d-tuples, and Y the space of m-tuples. We still have some freedom in specifying Ω, G, and Γ as spaces of functions defined on S. Notice we could require that elements of G be 0 at T, i.e., the system objective could be treated either as a constraint or as part of the definition of G.

Similarly, for the cost functional J we could take its domain to be $\Omega \times \Gamma$ or a set E of ordered pairs (u, y), where u is in Ω and $y = \gamma h$ for h satisfying the equation $h(t) = f(t) + \int_{[0, t]} h(s)\, ds + \beta u(t)$, t in S. We might also approximate the terminal contraint by assigning a large penalty proportional to $|h(T)|$ as part of J. Thus the formulation is a subtle balance of choices for the function spaces, the constraints, and the functional.

2. Optimization Background

We will pose each optimal control problem as a constrained optimization problem [25]. Basic for such problems are the necessary conditions provided by the Lagrange Multiplier Theorem. Suppose that each of G_1 and G_2 is a linear topological space with G_2 normed and F is a function from an open subset D of G_1 into G_2. Let x be a member of D and y be a member of G_1. If $\lim_{\alpha \to \infty}(1/\alpha)[F(x + \alpha y) - F(x)]$ exists, it is called the Gateaux differential of F at x with increment y and is denoted by $\delta F(x; y)$. If the limit exists for each y in G_1, the function F is said to be Gateaux differentiable at x.

Suppose that each of G_1 and G_2 are normed by N_1 and N_2, respectively. The function F is said to be Frechet differentiable at x if there is a continuous linear transformation A from G_1 into G_2 such that

$$\lim_{N_1(y) \to 0} N_2(F(x + y) - F(x) - Ay)/N_1(y) = 0$$

Exercise. If F is Frechet differentiable at x then there is only one continuous linear transformation from G_1 into G_2 satisfying the definition. We will denote this continuous linear transformation by $F'(x)$.

Exercise. If F is Frechet differentiable at x and y is in G_1, then $F'(x)y$ is called the Frechet differential at x with increment y. If F is Frechet differentiable at x then F is Gateaux differentiable at x and $F'(x)y = \delta F(x; y)$ for each y in G_1.

Exercise. If F is Gateaux differentiable at x, F need not be Frechet differentiable at x.

Exercise. If F is Frechet differentiable at x then F is continuous at x.

If F is Frechet differentiable on an open set containing x and F' is continuous at x then F is said to be continuously Frechet differentiable at x. (F' is to be continuous as a function from an open subset of G_1 into the space of continuous linear transformations from G_1 into G_2 normed with the operator norm.)

Exercise. Suppose that F_1 is a transformation mapping an open set D in G_1 into an open set E in G_2 and F_2 is a transformation mapping E into a normed space $\{G_3, N_3\}$. Let $F = F_2F_1$. Suppose that F_1 is Frechet differentiable at x in D and F_2 is Frechet differentiable at $y = F_1(x)$ in E. Then F is Frechet differentiable at x and $F'(x) = F_2'(y)F_1'(x)$.

Exercise. Suppose that F is Frechet differentiable on an open domain D contained in G_1. Let x be in D, y be in G_1, and suppose that $x + \alpha y$ is in D for all α, $0 \le \alpha \le 1$, Then

$$N_2(F(x + y) - F(x)) \le N_1(y) \sup_{0\le\alpha\le 1} N_2(F'(x + \alpha y))$$

Suppose that each of $\{G_1, Q_1\}$ and $\{G_2, Q_2\}$ is a complete inner product space. Let $N_i = Q_i(\cdot, \cdot)^{1/2}$, for $i = 1, 2$.

Theorem 2.1. (Generalized Inverse Function Theorem.) Suppose that F is a continuously Frechet differentiable function from G_1 into G_2 with a regular point x_0 (i.e., $F'(x_0)$ maps G_1 onto G_2). There is an ordered pair (r, b) of positive numbers with the property that if y is in G_2 and $N_2(y - F(x_0)) < r$ then there is an x in G_1 such that $y = F(x)$ and $N_1(x - x_0) \le bN_2(y - F(x_0))$.

If $\{G, Q\}$ is a complete inner product space, f is a function from G to the scalars, and f is Frechet differentiable at x_0 in G, then the gradient of f at x_0, denoted by $\nabla f(x_0)$, is that member y of G such that $f'(x_0)x = Q(x, y)$, for all x in G. Let D denote the subset of G to which x belongs only in case $f(x) = f(x_0)$. Let E denote the subset of G to which x belongs only in case $x - x_0$ is orthogonal to $\nabla f(x_0)$. We will say that E is tangent to D at x_0.

Exercise. Suppose that $G = R^2$ and Q is the usual Euclidean inner product. Let f denote the function from G to R defined by $f(x) = u^2 - v$, for $x = (u, v)$ in G. Find the gradient of f at (1, 1).

Suppose that each of $\{G_1, Q_1\}$ and $\{G_2, Q_2\}$ is a complete inner product space, f is a function from G_1 to the reals, and F is a function from G_1 into G_2.

Exercise. If F is Frechet differentiable at x_0 in G_1 and E is the subset of G_1 to which x belongs only in case $x - x_0$ is in $N(F'(x_0))$, then E is tangent to the level set $D = \{x \text{ in } G_1 \mid F(x) = F(x_0)\}$ in the following sense. A member x of G_1 is in E only in case, for each g in G_2, x is in the subset of G_1 tangent to the level set $\{y \text{ in } G_1 | Q_2(F(y) - F(x_0), g) = 0\}$ at x_0.

Exercise. Suppose that $G_1 = R^3$ and $G_2 = R^2$ with Q_i the usual Euclidean inner product for i = 1, 2. Let F denote the function from G_1 into G_2 defined by $F(x) = (u^2 + v^2 + w^2 - 1, u^2 + v^2 - 1/2)$, for $x = (u, v, w)$ in G_1. Find the subset E of G_1 which is tangent to the level set $D = \{x \text{ in } G_1 | F(x) = 0\}$ at $(1/2, 1/2, 1/2^{1/2})$.

Theorem 2.2. (The Lagrange Multiplier Theorem) Suppose that x_0 is a local extremum of f on $D = \{x \text{ in } G_1 | F(x) = 0\}$, each of f and F is continuously Frechet differentiable in an open subset of G_1 containing x_0, and x_0 is a regular point of F. There is a point y of G_2 such that the Lagrangian functional defined by $L(x) = f(x) + Q_2(F(x), y)$, for x in G_1, is stationary at x_0, i.e., $L'(x_0)x = f'(x_0)x + Q_2(F'(x_0)x, y) = 0$ for each x in G_1.

Lemma 2.3. The gradient $\nabla f(x_0)$ is in $N(F'(x_0))^{\perp}$.

Corollary 2.4. Assuming the hypothesis of the Theorem with the exception that the range of $F'(x_0)$ is closed rather than G_2, there is a nonzero element (r, y) of $R \times G_2$ such that the functional defined by $L(x) = rf(x) + Q_2(F(x), y)$, for x in G_1, is stationary at x_0.

Exercise. Find the triangle with the largest area given the length on one side and the perimeter.

Example. In the "standard" setting, let $S = [0, 1]$. Suppose that B is in $\boldsymbol{B}$ and c is a positive number such that $N_w(P_vBf - P_uBf) \le c\int_{[u,v]}v\,|\,f\,|\,dI$, for each f in G and $u \le v \le w$ in S. Let $k = 1 + I$ and J denote the function from G_H to the numbers defined by $J(f) = (1/2)N_H(f + Bf)^2$. Suppose that (x, y) is in X^2. Let A denote the function from G_H into X^2 defined by $A(f) = (f(0) - x, f(1) - y)$. We want to find a local minimum of J on $D = \{f \text{ in } G_H \mid Af = 0\}$.

Note that J has a unique minimum on D. If (f, g) is in $G_H \times G_H$ then $J'(f)g = Q_H(g + Bg, f + Bf)$ and $A'(f)g = (g(0), g(1))$. If f is the minimum of J in D then by the Lagrange Multiplier Theorem there is a (λ_1, λ_2) in X^2 such that $Q_H(g + Bg, f + Bf) + \langle g(0), \lambda_1\rangle + \langle g(1), \lambda_2\rangle = 0$ for each g in G_H.

Hence $Q_H(g, (I + B^*)(I + B)f) + Q_H(g, K(\ , 0)\lambda_1) + Q_H(g, K(\ , 1)\lambda_2) = 0$ for each g in G_H or $(I + B^*)(I + B)f + \lambda_1 + K(\ , 1)\lambda_2 = 0$.

Suppose that f is in G_H, (λ_1, λ_2) is in X^2, $f(0) = x$, $(I + B^*)(I + B)f + \lambda_1 + K(\ , 1)\lambda_2 = 0$, and $f(1) = z$. If g is in G_H, $g(0) = x$, and $g(1) = z$, then

$$\begin{aligned} J(f) &= (1/2)\, Q_H(f + Bf, f + Bf) \\ &= (1/2)\, Q_H(f, (I + B^*)(I + B)f) \\ &= (1/2)\, Q_H(f, -K(\ , 0)\lambda_1 - K(\ , 1)\lambda_2) \\ &= -(1/2)\, \langle f(0), \lambda_1\rangle - (1/2)\, \langle f(1), \lambda_2\rangle \\ &= -(1/2)\, \langle g(0), \lambda_1\rangle - (1/2)\langle g(1), \lambda_2\rangle \\ &= -(1/2)\, Q_H\, (g, K(\ , 0)\lambda_1 + K(\ , 1)\lambda_2) \\ &= (1/2)\, Q_H\, ((I + B)g, (I + B)f) \end{aligned}$$

Thus

$$\begin{aligned} 0 &\le (1/2)\, Q_H((I + B)(f - g), (I + B)(f - g)) \\ &= (1/2)\, Q_H((I + B)f, (I + B)f) - Q_H((I + B)f, (I + B)g) + \\ &\qquad (1/2)\, Q_H((I + B)g, (I + B)g) \\ &= (1/2)\, Q_H((I + B)g, (I + B)g) - (1/2)\, Q_H((I + Bf, (I + B)f) \\ &= J(g) - J(f) \end{aligned}$$

i.e., $J(f) \le J(g)$.

If f is in G_H, (λ_1, λ_2) is in X^2, $f(0) = x$, $(I + B^*)(I + B)f + \lambda + K(\ , 1)\lambda_2 = 0$, and $f(1) = z$ then

$$f = -\lambda_1 + K(, 1)\lambda_2 - B^*f - Bf - B^*Bf$$
$$= -\lambda_1 - K(, 1)(\lambda_1 + x) + Cf$$

where $Cg = -Bg - B^*g + K(, 1)[B^*g](0) - B^*Bg + K(, 1)[B^*Bg](0)$ for each g in G_H. On the other hand, if f is in G_H, λ_1 is in X, $f = -\lambda_1 + K(, 1)(\lambda_1 + x) + Cf$, and $f(1) = z$, then $f(0) = x$ and $(I + B^*)(I + B)f + \lambda_1 + K(, 1)\lambda_2 = 0$, where $\lambda_2 = x - \lambda_1 - [B^*Bf](0) - [B^*f](0)$.

Thus finding the local extremum of J on D is equivalent to finding a λ_1 in X so that if $f = - \lambda_1 + K(, 1)(\lambda_1 + x) + Cf$ then $f(1) = y$. We have used the Lagrange Multiplier Theorem to rewrite the original problem as a boundary value problem. The solution of the boundary value problem is

$$\lambda_1 = \{[(I - C)^{-1} K(, 1)](1) - [(I - C))^{-1}K(, 0)](1)\}^{-1} \{y - [(I - C)^{-1} K(, 1)](1)x\}.$$

We know that $[(I - C)^{-1}K(, 1)(1) - [(I - C)^{-1}K(, 0)](1)$ must have an inverse because the original problem has a unique solution and hence the boundary value problem must have a unique solution.

We are now ready to take up some optimal control problems. We begin with a simplified problem for state space systems elaborating the problem later by adding an input process and an observation process. Turning to the problem for hereditary systems we follow a similar plan but do not achieve quite the same level of completeness.

3. The Basic Optimal Control Problem

Our major concern in these notes is system problems for hereditary systems. However, in order to make connections with the theory which is applicable to state space systems, we will take up a couple of "standard" problems of linear quadratic control [3]. The state space theory is much more extensive than we can indicate here. For instance, we treat the important problem of feedback stabilization only in passing.

Recall our assumption that $\{X, | \cdot |\}$ is a complete inner product space, r is a nonnegative number, and S is a closed subinterval of $[-r, \infty)$ containing -r and 0. Further, our continuing assumption is that $|f(u) - f(v)| \leq N_v(P_vf - P_uf)$ for $u \leq v$ in S and f in G. For the state space problems we will also assume that X is the space of d-tuples of scalars for d a positive integer, $<\cdot, \cdot>$ is the usual Euclidean inner product, and $r = 0$.

Suppose that $S = [0, T]$ for some positive number T, α is a $d \times d$ matrix, $B = \alpha C$ where C is the member of ***B*** defined by $[Cf](t) = \int_{[0,t]} f\, dI$ for each f in G and t in S, $A = (I - B)^{-1}$, $k = 1 + I$, and f is a fixed member of G_H. The first problem [35] is to

$$\begin{array}{ll} \text{Minimize on } (I - P_0)G_H \times G_H: & J(u, h) = 1/2\, N_H(u)^2 + 1/2\, N_H(Ch)^2 \\ \text{Subject to:} & h = f + u + Bh,\ h(T) = 0 \end{array}$$

If (u, h) solves the problem then the Lagrange Multiplier Theorem asserts the existence of an element (λ, μ) of $G_H \times X$ such that

$$Q_H(u',u) + Q_H(h', C^*Ch) + Q_H(h' - u' - Bh', \lambda) + Q_H(h', K(\ , T)\mu) = 0$$

for each (u', h') in $(I - P_0)\, G_H \times G_H$. Hence $u = (I - P_0)\lambda$ and $C^*Ch + (I - B^*)\lambda + K(\ , T)\mu = 0$. Eliminating μ we have $C^*Ch - k[C^*Ch](0) + \lambda - B^* \lambda - k(\lambda(0) - [B^* \lambda](0)) = 0$.

Let $B_1 = B^* - k[B^*\cdot](0)$. Then B_1 is in ***B*** and $\lambda = (I - B_1)^{-1} k\, \lambda(0) + (I - B_1)^{-1}C^2h$. Note that $C^2 = k[C^*C\cdot](0) - C^*C$. Since $h = f + (I - P_0)\lambda + Bh$ we have

$$\begin{aligned} h &= f + (I - P_0)(I - B_1)^{-1}k\lambda(0) + ((I - B_1)^{-1}C^2 + B)h \\ &= (I - (I - B_1)^{-1}C^2 - B)^{-1}(f + (I - P_0)(I - B_1)^{-1}k\lambda(0)) \end{aligned}$$

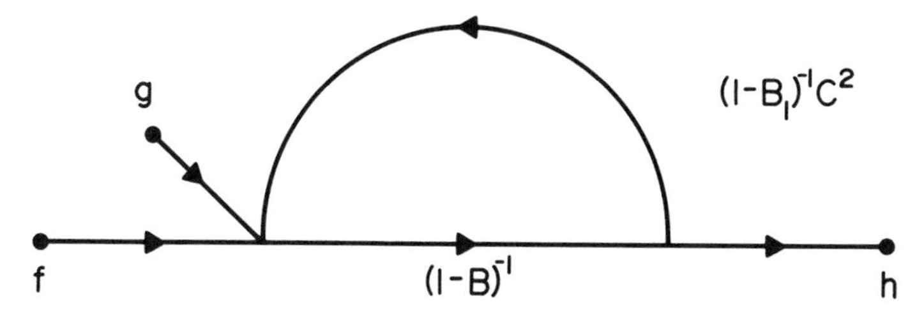

Figure 2.1

Since h(T) = 0,

$$\lambda(0) = -\{[\{I - (I - B_1)^{-1}C^2 - B\}^{-1}(I - P_0)(1 - B_1)^{-1}k](T)\}^{-1} [(I - (I - B_1)^{-1}C^2 - B)^{-1}f](T)$$

Let $g = (I - P_0)(1 - B_1)^{-1}k\lambda(0)$. Then $h = f + g + ((I - B_1)^{-1}C^2 + B)h$ or $u = g + (I - B_1)^{-1}C^2h$ which we can diagram as in Figure 2.1.

We see that u is the sum of an 'open loop' control g and a dynamic feedback $(I - B_1)^{-1}C^2h$.

Exercise. Show that if $f(t) = h(0)$ for $0 \leq t \leq T$ then the optimal control consists only of feedback.

Example. In order to illustrate the basic problem, consider the following simple example. Let $d = 1$, G be the standard example of continuous functions on $S = [0, 1]$, and $\alpha = 1$. If f is in G and t is in S then

$$[Bf](t) = \int_0^t f(s)\, ds$$

$$[B^*f](t) = (1 + t)f(1) - f(0) - \int_0^t f(s)\, ds$$

$$[B_1f](t) = \int_0^t (f(0) - f(s))\, ds$$

$$[C^2f](t) = \int_0^t (t - s)f(s)\, ds$$

Further, one can easily see that $[(I - B_1)^{-1}f](t) = f(0) + \int_{[0, t]} e^{-(t-s)}\, df(s)$ and $[(I - B_1)^{-1}C^2f](t) = \int_{[0, t]}(1 - e^{(t-s)})f(s)\, ds$. Using Laplace transforms one can show that

$$[(I - (I - B_1)^{-1}C^2 - B) f](t) = f(0)\{\cosh 2^{1/2} + (1/2^{1/2})\sinh(2^{1/2}t)\}^{-1}$$
$$\int_{[0,t]}\{\cosh 2^{1/2}(t - s) + (1/2^{1/2}) \sinh 2^{1/2}(t - s))\}\, df$$

To find $\lambda(0)$, note that $[(I - B_1)^{-1}k](t) = 1 + \int_{[0, t]} e^{-(t-s)}\, ds = 2 - e^{-t}$. Assume in the optimal control problem that $f(t) = f(0)$ for t in S. Then

$$\lambda(0) = -f(0)(\cosh 2^{1/2} + (1/2^{1/2})\sinh 2^{1/2}) / \int_0^1 \{\cosh 2^{1/2}(1 - s))$$
$$+ (1/2^{1/2})(1 - s))\}e^{-s}\, ds$$
$$\approx -2.5918f(0)$$

and $g(t) \approx 2.5918(e^{-t} - 1)f(0)$ for all t in S.

Alternate forms for the cost functional and constraints can be introduced to replace the terminal condition. For instance, instead of requiring $h(T) = 0$ one might simply apply a cost to $| h(T) |$ as in the following problem.

Minimize on $(I - P_0)\, G_H \times G_H$: $J(u, h) = 1/2\, N_H(u)^2 + 1/2\, N_H(Ch)^2 + b/2\, | h(T) |^2$

Subject to: $h = f + \alpha Ch + u$

Again, if (u, h) is a solution then there is a λ in G_H such that $u = (I - P_0)\lambda$, $\lambda = -A^*(C^*Ch + bK(, T)h(T))$ with $A = (I - \alpha C)^{-1}$, and $h = f + \alpha Ch + (I - P_0)\lambda$.

Exercise. Show that $(I - P_0)A^*C^*Ch = -(I + \alpha^*C)^{-1}C^2h + (I - P_0)(I + \alpha^*C)^{-1} K(, T)[Ch](T) + (I - P_0)(I + \alpha^*C)^{-1}\alpha^*K(, T)[A^*C^*Ch](T)$.

Hence

$$h = f + \alpha Ch - (I - P_0)A^*(C^*Ch + bK(, T)h(T))$$
$$= A_1(f - k_1h(T) - k_2[Ch](T) - k_3[A^*C^*Ch](T))$$

where $A_1 = (I - \alpha C - (I + \alpha^*C)^{-1}C^2)^{-1}$, $k_1 = b(I - P_0)\, A^*K(, T)$, $k_2 = (I - P_0)(I + \alpha^*C)^{-1}K(, T)$, and $k_3 = (I - P_0)(I + \alpha^*C)^{-1}\alpha^*K(, T)$. We solve the following relations for the vectors $x_1 = h(T)$, $x_2 = [Ch](T)$, and $x_3 = [A^*C^*Ch](T)$ to obtain representations of u and h.

$$x_1 = [A_1f](T) - [A_1k_1](T)x_1 - [A_1k_2](T)x_2 - [A_1k_3](T)x_3$$
$$x_2 = [CA_1f](T) - [CA_1k_1](T)x_1 - [CA_1k_2](T)x_2 - [CA_1k_3](T)x_3$$
$$x_3 = [A^*C^*CA_1f](T) - [A^*C^*CA_1k_1](T)x_1 - [A^*C^*CA_1k_2](T)x_2 - [A^*C^*CA_1k_3](T)x_3$$

Example. After some tedious computation we obtain for

$$\alpha = \begin{bmatrix} 4 & -2 \\ -2 & .1 \end{bmatrix}$$

and various b's some values for the x's.

For b = 1,

$$x_1 = \begin{bmatrix} -0.096231 \\ -0.022813 \end{bmatrix} \quad x_2 = \begin{bmatrix} 0.12232 \\ 0.2246 \end{bmatrix} \quad x_3 = \begin{bmatrix} -0.0246 \\ 0.23918 \end{bmatrix}$$

For b = 2,

$$x_1 = \begin{bmatrix} -0.047686 \\ -0.015249 \end{bmatrix} \quad x_2 = \begin{bmatrix} 0.13934 \\ 0.24101 \end{bmatrix} \quad x_3 = \begin{bmatrix} -0.008516 \\ 0.24198 \end{bmatrix}$$

For b = 10,

$$x_1 = \begin{bmatrix} -0.010531 \\ -0.0053776 \end{bmatrix} \quad x_2 = \begin{bmatrix} 0.15074 \\ 0.25108 \end{bmatrix} \quad x_3 = \begin{bmatrix} -8.7905 \times 10^{-4} \\ 0.24384 \end{bmatrix}$$

For b = 200,

$$x_1 = \begin{bmatrix} -1.468 \times 10^{-4} \\ -7.1122 \times 10^{-5} \end{bmatrix} \quad x_2 = \begin{bmatrix} 0.15292 \\ 0.25235 \end{bmatrix} \quad x_3 = \begin{bmatrix} -0.0016446 \\ 0.24418 \end{bmatrix}$$

Herditary systems. When we turn to the problem of optimal control of hereditary systems we can avoid some troublesome difficulties by altering the basic problem. Instead of system descriptions of the form $h = f + Bh$ with B in $\boldsymbol{B}$, we will use the equivalent description $h = A(f + u)$ with $A = (I - B)^{-1}$ in $\boldsymbol{A}$. This change suggests functionals of the form $J(u, h) = (1/2)N_H(Au)^2 + (1/2)N_H(Ch)^2$.

Note that the control cost is computed 'indirectly', i.e., instead of assessing the control cost directly in terms of u we use the response to the control u of the system initially at 0. Altering the problem allows us to avoid computing B^* for a general B in $\boldsymbol{B}$. If we review the argument for the state space problem we note that $B_1 = B^* - k[B^*\cdot](0)$ is in $\boldsymbol{B}$. This does not hold in general for hereditary systems.

Exercise. Assume

$$[Bh](t) = \begin{cases} 0 & -r \leq t \leq 0 \\ \int_0^t h(s-r)\,ds & 0 < t \end{cases}$$

for h in G_H. Show that $B_1 = B^* - k[B^*\cdot](0)$ is not in ***B***.

Of course, starting with a formal delay differential or integral equation we may be able to introduce function spaces and operators to obtain an 'equivalent' state space formulation [2, 7, 12, 30]. The underlying space X would then of necessity be infinite dimensional, and solving the equation resulting from the terminal conditions offers some new difficulties. We must either change something or bog down in technical details. Our approach is to change the form of the functional J.

Assuming $S = [-r, T]$, k is an increasing function continuous on $[-r, 0)$ and $[0,T]$,

$$[Ch](t) = \begin{cases} 0 & -r \leq t \leq 0 \\ \int_0^t h\,dk & 0 \leq t \leq T \end{cases}$$

for h in G_H, and t in S, and f is in G_H the basic problem becomes

Minimize on $(I - P_0)G_H \times G_H$: $J(u, h) = 1/2\, N_H(Au)^2 + 1/2\, N_H(Ch)^2$
Subject to: $h = A(f + u),\ (I - P_{T-r})h = 0,\ h(T) = 0.$

Exercise. Show that $[C^*h](t) = K(t, T)\, h(T) - K(0, t)h(0) - [Ch](t)$ for $0 \leq t \leq T$ and h in G_H.

If (u, h) solves the problem then the Lagrange Multiplier Theorem asserts the existence of an element (κ, λ, μ) of $(I - P_{T-r})G_H \times G_H \times X$ such that $Q_H(Au', Au) + Q_H(h', C^*Ch) + Q_H(h', \kappa) + Q_H(h' - Au', \lambda) + Q_H(h', K(\ , T)\mu) = 0$ for each (u', h') in $(I - P_0)G_H \times G_H$. Hence $Au = (I - P_0)\lambda$ and $C^*Ch + \kappa + \lambda + K(\ , T)\mu = 0$. Eliminating μ, we have $C^*Ch - (1/k(0))k[C^*Ch](0) + \lambda - (1/k(0))k\lambda(0) + \kappa = 0$. Since $h = Af + (I - P_0)\lambda$ we have

$$\begin{aligned} h &= Af + C^2h + (1/k(0))(I - P_0)k\lambda(0) - \kappa \\ &= (I - C^2)^{-1} Af + (1/k(0))(I - C^2)^{-1}(I - P_0)k\lambda(0) - (I - C^2)^{-1}\kappa. \end{aligned}$$

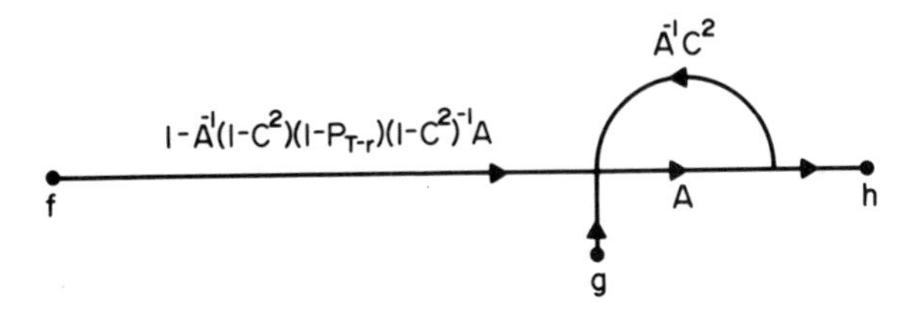

Figure 2.2

Let $k_1 = (1/k(0))(I - P_0)k$. Since $h(T - r) = 0$, $0 = [(I - C^2)^{-1}Af](T - r) + [(I - C^2)^{-1}k_1](T - r)\lambda(0)$ or $\lambda(0) = - \{[(I - C^2)^{-1}k_1](T - r)\}^{-1}[(I-C^2)^{-1}Af](T - r)$. Since $(I - P_{T-r})h = 0$ and $P_{T-r}\,\kappa = 0$, $\kappa = (I - C^2)(I - P_{T-r})(I - C^2)^{-1}\{Af + k_1\lambda(0)\}$, $\lambda = C^2h + (1/k(0))k\lambda(0) - \kappa$, and $u = (A^{-1}C^2h + A^{-1})k_1\lambda(0) - A^{-1}\kappa$.

Substituting for $\lambda(0)$ and κ we see that

$$u = A^{-1}C^2h - A^{-1}(I - C^2)(I - P_{T-r})(I - C^2)^{-1}Af$$
$$+ A^{-1}(I - (I - C^2)(I - P_{T-r})(I - C^2)^{-1})k_1\lambda(0).$$

Hence u is the sum of three terms, a dynamic feedback $A^{-1}C^2h$, a causal function $-A^{-1}(I - C^2)(I - P_{T-r})(I - C^2)^{-1}A$ of the input function f, and a function g which depends on some knowledge of f on the whole interval $[-r, T - r]$. We can diagram the controlled system as in Figure 2.2

Of course, if $r = 0$ then $A^{-1}(I - C^2)(I - P_{T-r})(I - C^2)^{-1}A = 0$.

Example. Let $d = 1$, G be the standard example of continuous functions on $S = [0, 1]$, $k = 1 + I$, and A the element of $\mathcal{A}$ defined by $[Af](t) = e^tf(0) + \int_{[0, t]} e^{t-s}\, df(s)$, for each f in G and t in S.

We have, for each f in G and t in S, $[(I - C^2)^{-1}f](t) = f(0)\cosh t + \int_{[0, t]}\cosh(t - s)\, df(s)$. Hence $[(I - C^2)^{-1}k_1](t) = \int_{[0, t]}\cosh(t - s)\, ds = -\sinh(t - s)\,|_{0, t} = \sinh t$. If $f(t) = f(0)$ for t in S then, for each t in S, $[Af](t) = e^tf(0)$ and

$$[(I - C^2)^{-1}AF](t) = f(0)\cosh t + \int_0^t \cosh(t - s)e^2\, ds\, f(0)$$

$$= (\cosh t + 2te^t + 2\sinh t)f(0)$$

Hence $\lambda(0) = -(7e - e^{-1})\, f(0)/(e - e^{-1})$ and $g(t) = (t - t^2/2)\lambda(0)$ for $t \geq 0$.

Response feedback stabilization. The problem of response feedback stabilization of linear, finite dimensional state, time invariant systems has many different solutions. Solution techniques have become so refined that one is concerned not so much with simple stabilization as with the "design" of a stabilizing feedback operator [35]. In this section we are interested in hereditary systems which do not have finite dimensional state space representation and which might not be time invariant [34, 38, 39, 40, 41].

Consider the system given formally by

$$h(t) = \begin{cases} f(t) & -r \leq t \leq 0 \\ f(t) + \int_0^t g(t, s)h(s)\, ds + \beta\int_0^t h(s - r)\, ds + u(t) & 0 \leq t \end{cases}$$

where u, h, and f could be vector valued and g and ß matrix valued. The basic problem is to find an operator D defined on some appropriate function space so that for some restricted class of f's the feedback control $u = Dh$ drives the response h to zero at infinity. A better problem is to choose D in terms of a desired decay rate for h.

Assume that $S = [-r, \infty)$, k is an increasing function continuous on $[-r, 0)$ and $[0, \infty)$, $\lim_{t\to\infty} k(t) = \infty$, and (A, B) with $A = (I - B)^{-1}$ is a fixed element of $\boldsymbol{A} \times \boldsymbol{B}$. We will assume that B has a variation function which is absolutely continuous with respect to k.

Exercise. Show that $\boldsymbol{B}$ maps G into G_H and hence A maps G_H onto G_H.

Finally, we want AP_TG_H to be contained in G_∞ for each $T \geq 0$. This can be achieved by requiring A - I to have a variation function k_1 such that $\int_{[-r, \infty]}(dk_1)^2/dk < \infty$.

Let f be a fixed element of G_H such that Af is in G_∞, U a closed linear subspace of X, and Ω the subset of $(I - P_0)G_H$ to which u belongs only in case u(S) is contained in U. In order to avoid discussing issues of controllability here, assume for each positive number T, $P_T A\Omega$ is dense in $(P_T - P_0)G_H$. The first optimization problem for fixed $T > 0$ can be stated as follows:

$$\text{Minimize on } \Omega \times G_H\text{:} \quad J(u, h) = 1/2\, N_H(Au)^2 + 1/2\, N_H(Ch)^2$$
$$\text{Subject to:} \quad h = A(f + u),\ \ h(t) = 0,\ \ T \le t.$$

Note that if (u, h) is a feasible solution then both Au and Ch are in G_∞.

We begin with a modification of this problem.

$$\text{Minimize on } \Omega \times G_H\text{:} \quad J_T(u, h) = 1/2\, N_H(P_T Au)^2 + 1/2\, N_H(P_T Ch)^2$$
$$\text{Subject to:} \quad h = P_T A(f + u),\ \ h(T) = 0.$$

<u>Exercise</u>. Show that $[(P_T C)^* h](t) = K(t, T)h(T) - K(0, t)h(0) - [P_T Ch](t)$ for $0 \le t$ and h in G_H.

If (u, h) solves this optimization problem then the Lagrange Multiplier Theorem asserts the existence of a pair (λ, μ) in $G_H \times X$ such that

$$Q_H(Au', P_T Au) + Q_H(h', (P_T C)^* P_T Ch) + Q_H(h' - Au', P_T \lambda)$$
$$+ Q_H(h', K(\ , T)\mu) = 0$$

for each (u', h') in $\Omega \times G_H$. But then $P_T Au = (P_T - P_0)\lambda$ and $(P_T C)^* P_T Ch + P_T \lambda + K(\ , T)\mu = 0$.

If $T \le t$ then $[(P_T C)^* P_T Ch](t) = [(P_T C)^* P_T Ch](T)$. If $t \le T$ then $[(P_T C)^* P_T Ch](t) = k(t)[Ch](T) - [C^2 h](t)$. Thus $[(P_T C)^* P_T Ch](0) = k(0)[Ch](T)$ and $\mu = -[Ch](T) - (1/k(0))\lambda(0)$.

So $P_T \lambda = P_T C^2 h + (1/k(0))\, P_T\, k\lambda(0)$ and $h = P_T Af + P_T C^2 h + P_T k'\lambda(0)$, where $k' = (1/k(0))(I - P_0)k$, or

$$h = P_T (I - C^2)^{-1}(Af + k'\lambda(0)) \qquad 2.1$$

where $\lambda(0) = -\,([(I - C^2)^{-1}k'](T))^{-1}[(I - C^2)^{-1}Af](T)$ and $Au = P_T C^2 h + P_T k'\lambda(0)$.

Looking back at the first problem, the only modification we need to make is to take

$$Au = P_T C^2 h + P_T k'\lambda(0) - (I - P_T)Af \qquad 2.2$$

Suppose (u', h') is a feasible solution of the first problem, (u, h) is defined by equations 2.1-2, and $\lambda = P_T C^2 h + (1/k(0))P_T k\, \lambda(0)$. Then

$$\begin{aligned} &Q_H(Ch', Ch) + Q_H(Au', Au) \\ &\quad = -Q_H(h', \lambda) + Q_H(h' - Af, (P_T - P_0)\lambda - (I - P_T)Af) \\ &\quad = -Q_H(h', \lambda) + Q_H(h', \lambda) - Q_H(Af, \lambda) + Q_H(Af, (I - P_T)Af) \\ &\quad = -Q_H(Af, \lambda - (I - P_T)Af) \\ &\quad = N_H(Ch)^2 + N_H(Au)^2 \end{aligned}$$

Hence

$$\begin{aligned} &N_H(A(u' - u))^2 + N_H(C(h' - h))^2 \\ &\quad = N_H(Au')^2 - 2Q_H(Au', Au) + N_H(Au)^2 + N_H(Ch)^2 \\ &\qquad\quad - 2Q_H(Ch', Ch) + N_H(Ch)^2 \\ &\quad = 2J(u', h') - 2J(u, h) \end{aligned}$$

i.e., (u, h) is optimal.

Theorem 2.5. Let $\{T_i\}_{i=0,\infty}$ be an increasing unbounded sequence in $(0, \infty)$ and, for $i = 0, 1, 2, \ldots$, let (u_i, h_i) be a solution of the first optimization problem with $T = T_i$; then $\{h_i\}_{i=0,\infty}$ and $\{Ch_i\}_{i=0,\infty}$ converge with respect to N_H. If h' is the limit of $\{h_i\}_{i=0,\infty}$ then both h' and Ch' are in G_∞ and so $\lim_{T\to\infty} h'(T) = 0$.

Proof. Repeating the previous sufficiency argument, if $i < j$ then $N_H(h_i - h_j)^2 + N_H(C(h_i - h_j))^2 = 2J(u_i, h_i) - 2J(u_j, h_j)$. Thus $\{h_i\}_{i=0,\infty}$ and $\{Ch_i\}_{i=0,\infty}$ converge with respect to N_H.

Let g be the limit of $\{Ch_i\}_{i=0,\infty}$. If T is in S then $\{P_T Ch_i\}_{i=0,\infty}$ converges to $P_T g = P_T Ch'$. Thus $N_H(P_T Ch') = N_H(P_T g) \le N_H(g)$, i.e., Ch' is in G_∞. Note that this implies $\lim_{T\to\infty} h'(T) = 0$. □

Exercise. Let h' and Ch' belong to G_∞. Let $\{T_p\}_{p=\infty}$ be an increasing unbounded sequence in S. Show that $\Sigma_{p=0,\infty} | h'(T_p) |^2 dk(T_p, T_{p+1}) \le \int_{[0,\infty]} | dCh' |^2 /dk$. Conclude that $\lim_{p\to\infty} | h'(T_p) |^2 dk(T_p, T_{p+1}) = 0$ and $\lim_{t\to\infty} h'(t) = 0$.

The response h' represents the trajectory we wish the 'stabilized' system to follow. The remaining problem is to find an element D of ***B*** such that $h' = f + (B + D)h'$, where $B = I - A^{-1}$. We will assume from this point that A_0 is a fixed element of ***A*** such that $A_0 P_0 G_H$ is contained in G_∞ and $f = A_0 P_0 f$. For i = 0,1,2, ···, let $D_i = (I - B)P_{T(i)}(C^2 + B_i)$, where

$$[B_i g](t) = - k'(t)\{[(I - C^2)^{-1}k'](T_i)\}^{-1}[(I - C^2)^{-1} AA_0P_0 g](T_i)$$

for each g in G_H and t in S. Note that D_i is in ***B*** for i = 0, 1, 2, ...

Theorem 2.6. If $\{B_i\}_{i=0,\infty}$ converges with respect to the operator norm N_T for each T in S then so does $\{D_i\}_{i=0,\infty}$. If D is the limit of $\{D_i\}_{i=0,\infty}$ with respect to N_T for each T in S then $h' = f + (B + D)h'$.

Proof. The first part of the theorem is immediate. Suppose that $h = f + (B + D)h$. If $T \le T_i$ then $P_T(h - h_i) = P_T C^2(h - h_i) + P_T k'(\lambda(0) - \lambda_i(0))$, where $\lambda_i(0) = [B_i f](T_i) / (k'(T_i) - k'(0))$ and $\lambda(0) = \lim_{i\to\infty}\lambda_i(0)$. Hence there is a number c such that $N_T(h - h_i) \le c \,|\, \lambda(0) - \lambda_i(0) \,|$, for i = 0, 1, 2, ···. Note that this implies $h = h'$.
□

Because of Theorem 2.6 we are interested in obtaining a representation of $(I - C^2)^{-1}$.

Theorem 2.7. For each $d \times d$ nonegative matrix α, f in G_h, and $t \ge 0$

$$[(I - \alpha C^2)^{-1} f](t) = f(t) - \int_0^t d\cosh(\alpha^{1/2}k(t) - \alpha^{1/2}k(s))f(s)$$
$$= \cosh(\alpha^{1/2}k(t) - \alpha^{1/2}k(0))f(0) + \int_0^t \cosh(\alpha^{1/2}k(t) - \alpha^{1/2}k(s))\, df(s)$$

Lemma 2.8. For each nonnegative integer m and subinterval [s, t] of S

$$\int_s^t (k(t) - k(x))^m \, d(k(t) - k(x)) = -(k(t) - k(s))^{m+1}/(m + 1)$$

Proof. The equality is obvious for m = 0. Further,

$$\int_s^t (k(t) - k(x))\, d(k(t) - k(x)) = -(k(t) - k(s))^2 - \int_s^t (k(t) - k(x))\, d(k(t) - k(x))$$
$$= -\, 1/2(k(t) - k(s))^2$$

Assuming the equality for m,

$$\int_s^t (k(t) - k(x))^{m+1} \, d(k(t) - k(x))$$
$$= -(k(t) - k(s))^{m+2}/(m + 1) - \int_s^t (k(t) - k(x))^{m+1} \, d(k(t) - k(x))/(m + 1)$$
$$= -(k(t) - k(s))^{m+2}/(m+2) \quad \square$$

Lemma 2.9. For $m = 0, \cdots, n$, $[C^{n+1}f](t) = -\int_{[0,\,t]}[C^{n-m}f](s)\, d(k(t) - k(s))^{m+1}/(m + 1)!$.

Proof. The equality obviously holds for m = 0. Assume it holds for some m, $0 \le m < n$.

$$[C^{n+1}f](t) = \frac{-1}{(m+1)}\int_0^t [C^{n-m}f](s)\, d(k(t) - k(s))^{m+1}$$
$$= \frac{1}{(m+1)!}\int_0^t (k(t) - k(s))^{m+1}[C^{n-m-1}f](s)\, dk(s)$$
$$= \frac{1}{(m+1)!}\int_0^t [C^{n-m-1}f](s)\, d\int_s^t (k(t) - k(x))^{m+1}\, d(k(t) - k(x))$$
$$= \frac{-1}{(m+2)!}\int_0^t [C^{n-m-1}f](s)\, d(k(t) - k(s))^{m+2} \qquad \square$$

Proof of Theorem 2.7. For each f in G_H and $t \ge 0$,

$$\sum_{p=1}^{\infty} \alpha^p [C^{2p} f](t) = f(t) - \int_0^t d[\sum_{p=1}^{\infty} \frac{\alpha^p}{(2p)!} (k(t) - k(s))^{2p} f(s)]$$

$$= f(t) - \int_0^t d[\cosh(\alpha^{1/2} k(t) - \alpha^{1/2} k(s)) f(s)]$$

$$= \cosh(\alpha^{1/2} k(t) - \alpha^{1/2} k(0) f(0) + \int_0^t \cosh(\alpha^{1/2} k(t) - \alpha^{1/2} k(s))\, df(s)$$

$$= [(I - \alpha C^2)^{-1} f](t) \quad \square$$

Theorem 2.10. There is an element D of $\boldsymbol{B}$ such that if f is in $A_0P_0G_H$ and $h = f + (B + D)h$ then $\lim_{t\to\infty} h(t) = 0$. Furthermore, Ch is in G_∞.

Proof. Using Theorem 2.6 we just have to show that for each f in $A_0P_0G_H$ $([(I-C^2)^{-1}k'](T))^{-1}[(I-C^2)^{-1}Af](T)$ has a limit as T increases without bound. We have

$$([(I - C^2)^{-1}k'](T))^{-1}[(I - C^2)^{-1}Af](T) = k(0)\{\sinh(k(T) - k(0))\}^{-1} \{\cosh(k(T) - k(0))f(0) + \int_0^T \cosh(k(T) - k(s))\, d[Af](s)\}$$

is asymptotic to $k(0)\{f(0) + \int_{[0,T]} \exp(-k(s))\, d[Af](s)\}$. Since Af is in G_∞, if $\{s_p\}_{p=0,n}$ is an increasing sequence in $[0, \infty)$ then

$$\sum_{p=1}^{n} \exp(-k(s_{p-1}) \mid [Af](s_p) - [Af](s_{p-1}) \mid$$

$$\leq \sum_{p=1}^{n} \exp(-k(s_{p-1}))[\int_{s(p-1)}^{s(p)} \mid dAf \mid^2 /dk]^{1/2} (k(s_p) - k(s_{p-1})^{1/2}$$

$$\leq N_H(Af)\{\sum_{p=1}^{n} \exp(-2k(s_{p-1}))\, dk(s_{p-1}, s_p)\}^{1/2}$$

i.e.,

$$\mid \int_a^b e^{-k(s)}\, d[Af](s) \mid \leq N_H(Af)\{ \int_a^b e^{-2k(s)}\, dk(s)\}^{1/2} = (1/2^{1/2}) N_H(Af)(e^{-2(k(b)-k(a))})^{1/2}$$

for each subinterval [a, b] of $[0, \infty]$. Hence the required limit exists and we are through. □

The significance of the fact that Ch is in G_∞, besides its use in the proof that $\lim_{t\to\infty} h(t) = 0$, is that we have an estimate on the rate of decay of the response. In fact, we can increase the decay rate by choosing a larger k. We will illustrate with an example.

Consider the scalar system given formally by the equation $h(t) = f(t) + \alpha\int_{[0,t]} h(s)\, ds$, where α is a positive number. The operator B is defined by $[Bh](t) = \alpha\int_{[0,t]} h(s)\, ds$ for each h in G and t in $S = [0, \infty)$. The operator A is defined by $[Af](t) = e^{\alpha t} f(0) + \int_{[0,t]} e^{\alpha(t-s)}\, df(s)$ for each f in G and t in S. We can take αI as a variation for B and $\exp(\alpha I)$ as a variation for A - I. The requirement that $AP_T f$ be in G_∞ for each T in S and f in G_H will be met by taking $k(t) = e^{3\alpha t}$ for each t in S. Let $A_0 = I$. The operator D is given by $Dh = (I - B)(C^2 h - k'\lambda(0))$ for h in G_H and t in S, where $k'(t) = e^{3\alpha t} - 1$ and $\lambda(0) = -(1 + \int_{[0,\infty]} e^{-(k(s)-k(0))}\, de^{\alpha s})h(0)$.

This scheme for choosing k works in general. If the system is given by (A, B) and k_1 is a variation function for B then $\exp(k_1)$ is a variation function for A - I and we can choose $k = \exp(3k_1)$. Furthermore, $[Dh](t) = (I - B)(C^2 h - k'\lambda(0))$, where $k' = (I - P_0)k$ and $\lambda(0) = -\{h(0) + \int_{[0,\infty]} e^{-(k(s)-k(0))} d[AA_0P_0h](s)\}$.

It is of interest that other choices for k might be available. Consider the scalar system $h(t) - f(t) - \int_{[0,t]} (t - s)h(s)\, ds$. Here $[Bh](t) = -\int_{[0,t]} (t-s)h(s)\, ds$ and $[Af](t) = (\cos t)f(0) + \int_{[0,t]} \cos(t - s)\, df(s)$ for h and f in G and t in $S = [0, \infty]$. For this example we can take $k(t) = 1 + t^3$ for t in S so that $\lambda(0)$ in the definition of Dh is given by $\lambda(0) = -(1 + \int_{[0,\infty]} \exp(-t^3)\, d \cos t)h(0)$ when $A_0 = I$.

The stabilization methods presented here apply to a rather general class of linear hereditary systems including Stieltjes-Volterra integral equations and functional differential equations. There is no requirement of time invariance, although that is the simplest case to deal with computationally. The methods are robust with regard to system representation. For instance, a state space representation is not required. If the system is given in terms of a B from $\boldsymbol{B}$ then the most difficult part of finding the feedback operator D is computing $(I - B)^{-1}A_0P_0$. This task can be simplified by restricting ones attention to f's from suitable finite dimensional subspaces of P_0G_H.

The method gives some control of the rate of decay of the system response h. Since both h and Ch are in G_∞ the response can be made to decay faster by choosing a larger k. This phenomenon requires more study, but there is the possibility of using k as a design parameter.

RHK spaces suggest natural finite dimensional approximations for the stabilizing feedback operators. The study of these approximations is undertaken in Chapter 3; however, our limited computational experience with the numerical implementation of these methods indicates that they are practical.

4. Elaboration of the Basic Problem

We can add significantly to the realism of our system control problems by adding the elements of an input process and an observation process. In general, we cannot expect to affect or observe directly all of the internal states of the system. The task then is to understand the relations among the input process, the system dynamics, and the observation process required for us to solve a control problem.

Recall that the set of inputs or controls for an hereditary system is assumed to be a linear space Ω of functions from S into U, a subspace of X. For simplicity, we will assume that Ω is contained in $(I - P_0)G_H$. We will assume that ρ, the input operator from Ω into G, has the following properties:

1) ρ is linear and causal
2) For each t in S, $P_t\rho$ is a continuous function from $P_t\Omega$ into P_tG_H, i.e., continuous with respect to N_H.

We are concerned with a finite interval, terminal constraint control problem. Assume that $S = [-r, T]$, A is in $\mathcal{A}$, and f is in G_H. Consider the following optimal control problem:

$$\text{Minimize on } (I - P_0)G_H \times G_H\text{:}\quad J(u, h) = 1/2\, N_H(Au)^2 + 1/2\, N_H(Ch)^2$$
$$\text{Subject to:}\quad h = A(f + u),\ (I - P_{T-r})h = 0,\ h(T) = 0.$$

<u>Exercise</u>. Show that $[C^*h](t) = K(t, T)\, h(T) - K(0, t)h(0) - [Ch](t)$ for $0 \le t \le T$ and h in G_H.

If (u, h) solves the problem then the Lagrange Multiplier Theorem asserts the existence of an element (κ, λ, μ) of $(I - P_{T-r})G_H \times G_H \times X$ such that $Q_H(Au', Au)$

$+ Q_H(h', C^*Ch) + Q_H(h', \kappa) + Q_H(h' - Au', \lambda) + Q_H(h', K(, T)\mu) = 0$ for each (u', h') in $(I - P_0)G_H \times G_H$. Hence $Au = (I - P_0)\lambda$ and $C^*Ch + \kappa + \lambda + K(, T)\mu = 0$. Eliminating μ, we have $C^*Ch - (1/k(0))k[C^*Ch](0) + \lambda - (1/k(0))k\lambda(0) + \kappa = 0$. Since $h = Af + (I - P_0)\lambda$ we have

$$\begin{aligned} h &= Af + C^2h + (1/k(0))(I - P_0)k\lambda(0) - \kappa \\ &= (I - C^2)^{-1} Af + (1/k(0))(I - C^2)^{-1}(I - P_0)k\lambda(0) - (I - C^2)^{-1}\kappa. \end{aligned}$$

Let $k_1 = (1/k(0))(I - P_0)k$. Since $h(T - r) = 0$, $0 = [(I - C^2)^{-1}Af](T - r) + [(I - C^2)^{-1}k_1](T - r)\lambda(0)$ or $\lambda(0) = - \{[(I - C^2)^{-1}k_1](T - r)\}^{-1}[(I-C^2)^{-1}Af](T - r)$. Since $(I - P_{T-r})h = 0$ and $P_{T-r}\, \kappa = 0$, $\kappa = (I - C^2)(I - P_{T-r})(I - C^2)^{-1}\{Af + k_1\lambda(0)\}$, $\lambda = C^2h + (1/k(0))k\lambda(0) - \kappa$, and $u = (A^{-1}C^2h + A^{-1})k_1\lambda(0) - A^{-1}\kappa$.

Substituting for $\lambda(0)$ and κ we see that

$$\begin{aligned} u = {} & A^{-1}C^2h - A^{-1}(I - C^2)(I - P_{T-r})(I - C^2)^{-1}Af \\ & + A^{-1}(I - (I - C^2)(I - P_{T-r})(I - C^2)^{-1})k_1\lambda(0). \end{aligned}$$

Hence u is the sum of three terms, a dynamic feedback $A^{-1}C^2h$, a causal function $-A^{-1}(I - C^2)(I - P_{T-r})(I - C^2)^{-1}A$ of the input function f, and a function g which cannot expect to affect or observe directly all of the internal states of the system. The task then is to understand the relations among the input process, the system dynamics, and the observation process required for us to solve a control problem.

Recall that the set of inputs or controls for an hereditary system is assumed to be a linear space Ω of functions from S into U, a subspace of X. For simplicity, we will assume that Ω is contained in $(I - P_0)G_H$. We will assume that ρ, the input operator from Ω into G, has the following properties:

1) ρ is linear and causal
2) For each t in S, $P_t\rho$ is a continuous function from $P_t\Omega$ into P_tG_H, i.e., continuous with respect to N_H.

We are concerned with a finite interval, terminal constraint control problem. Assume that $S = [-r, T]$, A is in $\mathcal{A}$, and f is in G_H. Consider the following optimal control problem:

$$\begin{aligned} &\text{Minimize on } \Omega \times G_H\text{:} && J(u, h) = 1/2\, N_H(A\rho u)^2 + 1/2\, N_H(Ch)^2 \\ &\text{Subject to:} && h = A(f + \rho u),\ h(t) = 0 \text{ for } T - r \le t \le T. \end{aligned}$$

In reviewing the solution for the special case of this problem considered in the previous section, we realize that the projection Π of $(I - P_0)G_H$ onto $cl(\boldsymbol{R}(A\rho))$ should have the following properties:

1) Π is causal and ΠC^2 is in $\boldsymbol{B}$.
2) $\Pi(I - P_{T-r}) = (I - P_{T-r})\,\Pi = I - P_{T-r}$
3) $[(I - \Pi C^2)^{-1}\,\Pi(I - P_0)K(\,,T)](T - r)$ is nonsingular.

Note that $\Pi = I - P_{r'}$ satisfies 1) - 3) if $r' < T - r$.

Exercise. Consider the system

$$h(t) = x_0 + \int_0^t \begin{bmatrix} 0 & 1 \\ -1 & 0 \end{bmatrix} h(s)\,ds + \begin{bmatrix} 0 \\ u(t) \end{bmatrix}$$

$$= \begin{bmatrix} \cos t & \sin t \\ -\sin t & \cos t \end{bmatrix} x_0 + \int_0^t \begin{bmatrix} \cos(t - s) & \sin(t - s) \\ -\sin(t - s) & \cos(t - s) \end{bmatrix} d\begin{bmatrix} 0 \\ u(s) \end{bmatrix}$$

Let $X = R^2$, $U = \{x \text{ in } X \mid x_1 = 0\}$, and $\rho = I$. Show that $R(A\rho) = \{\begin{bmatrix} f \\ g \end{bmatrix}$ in $(I - P_0)G_H \mid g = f'\}$ and that Π has the following representation:

$$[\Pi \begin{bmatrix} f \\ g \end{bmatrix}](t) = \begin{bmatrix} \int_0^t df(s)(1 - \cosh(t - s)) + \int_0^t dg(s)\sinh(t - s) \\ -\int_0^t df(s)\sinh(t - s) + \int_0^t dg(s)\cosh(t - s) \end{bmatrix}$$

Note that, $r = 0$, $S = [0, T]$, $\Pi(I - P_{T-r}) = (I - P_{T-r})\Pi = I - P_{T-r} = 0$, Π is causal, ΠC^2 is in $\boldsymbol{B}$ and, since $\Pi(I - P_0)G_H$ is invariant under C^2, $(I - \Pi C^2)^{-1} = I + (I - C^2)^{-1}\Pi C^2$. One can also show directly, though the computations are tedious, that $[(I - \Pi C^2)^{-1}\Pi(I - P_0)K(\,,T)](T - r)$ is nonsingular for $T > 0$.

Theorem 2.11. If conditions 1) - 3) hold for Π then for (u, h) in $(I - P_0)G_H \times G_H$ these are equivalent:

1) (u, h) is a solution of the optimal control problem;
2) $h = (I - \Pi C^2)^{-1}\{Af + \Pi(I - P_0)K(\,,T)\lambda'(0) - \upsilon\}$ and $A\rho u = \Pi(I - P_0)\lambda$, where

$\lambda'(0) = -\{[(I - \Pi C^2)^{-1}\Pi(I - P_0)K(\,,T)](T - r)\}^{-1}(I - \Pi C^2)^{-1}Af](T - r)$

$\upsilon = (I - \Pi C^2)(I - P_{T-r})(I - \Pi C^2)^{-1}(Af + \Pi(I - P_0)K(\,,T)\lambda'(0))$

$\lambda = C^2 h + K(\,,T)\lambda'(0) - \upsilon$

Proof. Suppose that 1) holds. The Lagrange Multiplier Theorem asserts the existence of an ordered triple (λ, μ, υ) in $G_H \times X \times (I - P_{T-r})G_H$ such that $Q_H(A\rho u', A\rho u) + Q_H(h', C^*Ch) + Q_H(h' - A\rho u', \lambda) + Q_H(h', K(\ ,T)\mu) + Q_H(h', \upsilon) = 0$, for all (u', h') in $(I - P_0)G_H \times G_H$. Therefore $A\rho u = \Pi(I - P_0)\lambda$ and $C^*Ch + \lambda + K(\ , T)\mu + \upsilon = 0$ or

$$\begin{aligned}\lambda &= C^2h + (1/k(0))K(\ ,T)\lambda(0) - \upsilon \\ &= C^2h + K(\ ,T)\lambda'(0) - \upsilon\end{aligned}$$

with $\lambda'(0) = \lambda(0)/k(0)$. It is helpful to note that $\Pi\upsilon = \Pi(I - P_{T-r})\upsilon = (I - P_{T-r})\upsilon = \upsilon$ and $(I - P_{T-r})(I - C^2)^{-1}\upsilon = (I - C^2)^{-1}\upsilon$. And so

$$\begin{aligned}h &= Af + A\rho u \\ &= Af + \Pi C^2h + \Pi(I - P_0)K(\ ,T)\lambda'(0) - \Pi\upsilon \\ &= (I - \Pi C^2)^{-1}\{Af + \Pi(I - P_0)K(\ ,T)\lambda'(0) - \upsilon\}\end{aligned}$$

Since $h(t) = 0$ for $T - r \leq t \leq T$,

$$\lambda'(0) = -\{[(I - \Pi C^2)^{-1}\,\Pi(I - P_0)K(\ ,T)](T - r)\}^{-1}[(I - \Pi C^2)^{-1}Af](T - r)$$

and

$$\upsilon = (I - \Pi C^2)(I - P_{T-r})(I - \Pi C^2)^{-1}\{Af + \Pi(I - P_0)K(\ ,T)\lambda'(0)\}.$$

Suppose that 2) holds. Clearly $h = A(f + \rho u)$ and $h(t) = 0$ for $T - r \leq t \leq T$, i.e., (u, h) is a feasible solution. If (u', h') is another feasible solution of the optimization problem then

$$\begin{aligned}Q_H(Ch', Ch) + Q_H(A\rho u', A\rho u) &= -Q_H(h', \lambda) + Q_H(h' - Af, \Pi(I - P_0)\lambda \\ &= -Q_H(Af, \lambda) \\ &= N_H(Ch)^2 + N_H(A\rho u)^2\end{aligned}$$

and so

$$N_H(A\rho(u' - u))^2 + N_H(C(h' - h))^2 = N_H(h' - h)^2 + N_H(C(h' - h))$$
$$= 2J(u', h') - 2J(u, h)$$

Hence $J(u', h') \geq J(u, h)$, i.e., (u, h) is a solution of the optimal control problem. □

Weak controllability of (A, ρ) in order to make use of the conditions on Π we need to look for some 'checkable' conditions on A and ρ which imply 1) - 3). This problem is only parially resolved here, i.e., we will assume that Π is causal and ΠC^2 is in $\boldsymbol{B}$ but develop conditions on A and ρ which imply 2) and 3). We say that the system (A, ρ) is weakly controllable on $[-r, T]$ provided

1) $[A\rho\Omega](T - r) = X$ and
2) $(I - P_{T-r})G_H$ is contained in $clA\rho\Omega$.

When $r = 0$, as would be the typical case for state space systems, the second condition is trivially satisfied and weak controllability becomes equivalent to the usual notion of controllability [35]. The exercise at the beginning of this section can be modified, take $r > 0$, and used to illustrate that 1) can hold but 2) not hold, i.e., weak controllability on $[-r, T]$ is stronger than controllability at T.

Exercise. Modify the example as indicated with $r > 0$ and show there is a nontrivial element of $(I - P_{T-r})G_H$ which is orthogonal to $clA\rho\Omega$.

For simplicity we assume that $S = [-r, T]$.

Theorem 2.12. If (A, ρ) is weakly controllable on $[-r, T]$ then

1) $\Pi(I - P_{T-r}) = (I - P_{T-r})\Pi = I - P_{T-r}$ and
2) $[(I - \Pi C^2)^{-1}\Pi(I - P_0)K(, T)](T - r)$ is nonsingular.

Proof. We have immediately that $\Pi(I - P_{T-r}) = (I - P_{T-r})$. If g is in G_H and f is in $A\rho\Omega$ then

$$Q_H((I - P_{T-r})\Pi(I - P_0)g, f) = Q_H((I - P_0)g, \Pi(I - P_{T-r})f)$$
$$= Q_H((I - P_{T-r})(I - P_0)g, f)$$

i.e., $(I - P_{T-r})\Pi = (I - P_{T-r})$.

If x is in X and $0 = \langle[\Pi(I - P_0)K(, T)x](T - r), x\rangle$

then

$$0 = Q_H(\Pi(I - P_0)K(, T)x, K(, T - r)\, x)$$
$$= N_H(\Pi(I - P_0)K(, T - r)x)^2$$

Thus $(I - P_0)K(, T - r)\, x$ is orthogonal to $A\rho\Omega$. If f is in $A\rho\Omega$ then

$$0 = Q_H(f, (I - P_0)K(, T - r)x)$$
$$= \langle f(T - r), x \rangle$$

Since $[A\rho\Omega](T - r) = X$, $x = 0$, i.e., $[\Pi(I - P_0)K(, T)](T - r)$ is nonsingular.

Suppose that x is in X and $[(I - \Pi C^2)^{-1}\Pi(I - P_0)K(, T)x](T - r) = 0$. Let $f = \Pi(I - P_0)K(, T - r)x$ and $h = (I - \Pi C^2)^{-1}f$. Then

$$\begin{aligned}
0 &= \langle [(I - \Pi C^2)^{-1}f](T - r), x \rangle \\
&= Q_H((I - \Pi C^2)^{-1}f, K(, T - r)x) \\
&= Q_H(P_{T-r}(I - \Pi C^2)^{-1}f, \Pi(I - P_0)K(, T - r)x) \\
&= Q_H(P_{T-r}h, (I - \Pi C^2)h) \\
&= N_H(P_{T-r}\, h)^2 - Q_H(\Pi P_{T-r}\, h, P_{T-r}\, C^2 h) \\
&= N_H(P_{T-r}\, h)^2 - \int_0^{T-r} \langle dh, Ch \rangle \\
&= N_H(P_{T-r}h)^2 - \langle h(T - r), [Ch](T - r) \rangle + \int_0^{T-r} |\, h\,|^2\, dI \\
&= N_H(P_{T-r}\, h)^2 + \int_0^{T-r} |\, h\,|^2\, dI
\end{aligned}$$

i.e., $0 = P_{T-r}\, h = P_{T-r}\, (I - \Pi C^2)^{-1}f$. Hence there is a $b > 0$ such that

$$\begin{aligned}
N_{T-r}(f) &= N_{T-r}((I - \Pi C^2)h) \\
&\le N_{T-r}(h) + b\, N_{T-r}(Ch) \\
&= 0\ . \qquad \Box
\end{aligned}$$

Thus $P_{T-r}\, f = 0$ and so $x = 0$. This means that $[(I - \Pi C^2)^{-1}\Pi(I - P_0K(, T)](T - r)$ is nonsingular.

Theorem 2.13. Let L denote the matrix representation of $A\rho$. These are equivalent:

1) $[A\rho\Omega](T - r) = X$
2) If $L(, T - r)x = 0$ then $x = 0$.

Exercise. Prove Theorem 2.13.

Exercise. Consider the variable coefficient finite dimensional state space system $h = h(0) + C(\alpha h) + \beta u$, where $\alpha = \begin{bmatrix} 1 & s \\ 0 & 1 \end{bmatrix}$, $\beta = \begin{bmatrix} 0 & 0 \\ 0 & 1 \end{bmatrix}$, and $S = [0, T]$. Show the system is weakly controllable.

Any function f on S of the form $f(t) = \Sigma_{p=0,n} K(t, s_p)x_p$ where $s = \{s_p\}_{p=0,n}$ is an increasing sequence in S and $\{x_p\}_{p=0,n}$ is a sequence of vectors in X is called a K-polygonal function. Clearly, any K-polygonal function is in G_∞ since each component is a linear combination of functions in G_∞. Also the subspace of all K-polygonal functions determined by a fixed partition s of [-r, T] is a closed linear subspace of G_∞. Let Π_s denote the orthogonal projection of G_∞ onto this subspace. The union of the finite dimensional subspaces $\Pi_s G_\infty$ is dense in $P_T G_H$ with respect to the Hellinger norm N_H. To see why this holds, suppose f is in $P_T G_H$ and is orthogonal to K(, t)x for each t in [-r, T] and x in X. Then $<f(t), x> = Q_H(f, K(, t)x) = 0$ and f is the zero function. The orthogonal complement of the closure of the union of the subspaces $\Pi_s G_\infty$ contains only the zero function, i.e., this union is dense in $P_T G_H$.

Theorem 2.14. These are equivalent:

1) $(I - P_{T-r})G_H$ is contained in $clA\rho\Omega$.

2) If the net $\{(A\rho)^*\Pi_s(I - P_{T-r})g$, s a partition of $[T - r, T]\}$ has limit 0 with respect to N_H then $(I - P_{T-r})g = 0$.

Exercise. Prove Theorem 2.14.

Problem. Translate the conditon in Theorem 2.14 as a condition on the matrix representation of $A\rho$.

Example. Consider the hereditary system example

$$h(t) = \begin{cases} f(t) & -1 \le t \le 0 \\ f(0) + \int_0^t \cos(t - s)h(s)\, ds + [\rho u[(t) & 0 \le t \end{cases}$$

Note that S = [-1, T], d = 1, and X = R. Assume that $\Omega = (I - P_0)G_H$. The input/output operator may be represented, for u in $(I - P_0)G_H$, as

$$[A\rho u](t) = 1/2 \int_0^t e^{1/2(t-\tau)}\{\sqrt{3}\sin\sqrt{3}/2(t - \tau) - \cos\sqrt{3}/2\,(t - \tau)\}d[\rho u](\tau)$$

Suppose that $[\rho u](t) = u(t - 1)$ for u in $(I - P_0)G_H$ and t in S. If g is in $(I - P_1)G_H$ and

$$u(t) = \begin{cases} 0 & -1 \le t \le 0 \\ -2g(t+1) + 4/5\int_0^t g(\tau+1)d\tau - 2\int_0^t e^{5/2(t-\tau)} g(\tau+1)d\tau & 0 \le t \end{cases}$$

Then $A\rho u = g$, i.e., $(I - P_1)G_H$ is contained in $A\rho(I - P_0)G_H$. If $T = 2$ then the first condition for weak controllability is not satisfied by the second holds.

Exercise. Show that if $T > 2$ then the system is weakly controllable.

Adding an observation process. The use of the prase 'observation process' suggests more than we want to take up in this subsection. Realism requires, having distinguished between internal and external parts of a system, that control be accomplished using observations rather than system responses. However, we will postpone the implied task of estimating the responses from the observations. Instead, we will undertake a much more modest problem, achieving control using observations ηh in place of h.

We assume that r' is a positive number less than r, $S = [-r, T + r']$, $k(t) = 1 + r + t$ for t in S, γ is a $d \times d$ matrix, and

$$[\eta g](t) = \begin{cases} \gamma g(-r) & -r \le t \le r' - r \\ \gamma g(t - r') & r' - r \le t \le T + r' \end{cases}$$

for each g in G. Recall that ηg has values in Y a subspace of X which has dimension less than d. In order to clarify the issues we will assume at first that $P_0 f$ is known. Consider the following optimal control problem.

$$\begin{array}{ll} \text{Minimize on } \Omega \times G_H\text{:} & J(u, h) = (1/2)\, N_H(A\rho u)^2 + (1/2)\, N_H((I - P_{r'})C\eta h)^2 \\ \text{Subject to:} & h = A(P_0 f + \rho u),\ [\eta h](t) = 0,\ T - r + r' \le t \le T + r'. \end{array}$$

Assuming for convenience that $\rho = I$, the Lagrange Multiplier Theorem asserts that if (u, h) is a solution then there exists (λ, μ, υ) in $G_H \times X \times (I - P_{T-r+r'})G_H$ such that $Q_H(Au', Au) + Q_H((I - P_{r'})C\eta h', (I - P_{r'})C\eta h) + Q_H(h' - Au', \lambda) + Q_H(\eta h', K(, T + r')\mu) + Q_H(\eta h', \upsilon) = 0$ for all (u', h') in $\Omega \times G_H$. Thus $Au = (I - P_0)\lambda$ and $(C\eta)^*(I - P_{r'})C\eta h + \lambda + \eta^* K(, T + r')\mu + \eta^*\upsilon = 0$.

One can show the following, for each g in G_H and t in S:

1) $[\eta^* g](t) = \gamma^* g(-r) + \gamma^*[(P_T - P_{-r})g(I + r')](t)$,

2) $[C^* g](t) = g(T + r')K(t, T + r')) - g(0)K(0, t) - [Cg](t)$,

3) $[\eta^* C^* g](t) = \gamma^*\{g(T + r') - g(0)\} + \gamma^* g(T + r')[(P_T - P_{-r})k(I + r')](t)$
$\quad -\gamma^* g(0)[(P_T - P_{-r})K(0, I + r')](t) - \gamma^*[P_T Cg](t + r')$

4) $[C\eta g](t) = \begin{cases} 0 & -r \le t \le 0 \\ \int_0^t \gamma g(s - r')\, dk(s) & 0 \le t \le T + r' \end{cases}$

5) $[(I - P_{r'})C\eta g](t) = \begin{cases} 0 & -r \le t \le r' \\ \int_0^{t-r'} \gamma g(s)\, dk(s) & r' \le t \le T + r' \end{cases}$

6) $[(C\eta)^*(I - P_{r'})C\eta g](t) = \gamma^*\gamma[Cg](T)[P_T k](t) - \gamma^*\gamma[P_T C^2 g](t)$

Therefore $[\eta^* K(, T + r')](t) = \gamma^*[P_T k](t)$, $[\eta^*\upsilon](t) = 0$ for $-r \le t \le T - r$, $k(0)\gamma^*\gamma[Ch](t) + k(0)\gamma^*\mu + \lambda(0) = 0$, $\lambda = \gamma^*\gamma P_T C^2 h + (1/k(0))P_T k\, \lambda(0) - \eta^*\upsilon$, and $h = AP_0 f + \gamma^*\gamma P_T C^2 h + k'\lambda(0) - \eta^*\upsilon$, where $k' = (1/k(0))(P_T - P_0)k$. Note that $\lambda(0)$ is in the range of γ^*.

Since $[\eta h](t) = 0$ for $T - r + r' \le t \le T + r'$, denoting the pseudo inverse of γ by $\gamma^\dagger$, we have $\gamma^\dagger\gamma h(t) = 0$ for $T - r \le t \le T$, $\lambda(0) = \gamma^\dagger\gamma\lambda(0) = ([I - \gamma^*\gamma C^2)^{-1}k'](T - r))^{-1}\,[(I - \gamma^*\gamma C^2)^{-1}\gamma^\dagger\gamma AP_0 f](T - r)$, $\eta^*\upsilon = (I - \gamma^*\gamma P_T C^2)(P_T - P_{T-r})(I - \gamma^*\gamma PC)^{-1}(\gamma^*\gamma AP_0 f + k'\lambda(0))$, $h = (I - \gamma^*\gamma P_T C^2)^{-1}(AP_0 f + k'\lambda(0) - \eta^*\upsilon)$, and $Au = (I - P_0)\lambda$. Note that $\lambda = \eta^\dagger\eta\gamma$, i.e., u depends on ηh rather than just h. One can show that if $\lambda(0)$, $\eta^*\upsilon$, h, λ, and u are defined this way then (u, h) solves the optimization problem.

Exercise. Assuming $P_0 f = 0$ and z is in γX solve the following optimal control problem.

Minimize on $\Omega \times G_H$: $J(u, h) = 1/2\, N_H(Au)^2 + 1/2 N_H(C\eta h)^2$
Subject to: $h = Au$; $[\eta h](t) = z$, $T - r + r' \le t \le T + r'$.

Reviewing the solution, one realizes that two difficulties remain. First, we need to deal with the assumption that $P_0 f$ is known at $t = 0$. Secondly, we need to investigate the implications for h of the constraint $[\eta h](t) = 0$ for $T - r + r' \le t \le T + r'$. One might feel that the first assumption is as unrealistic as the use of the system response in feedback control. We can partially overcome this objection by restricting the f's to a linear subspace G_0 of G_H which can be determined at time $t = 0$ from observations ηf on $[-r, 0]$.

For instance, we might assume the existence of a known continuous linear function A_0 from $P_{-r}G_H$ into G_H and take $G_0 = A_0 P_{-r} G_H$. The matrix representation L_0 of A_0 has the property that $L_0(s, t) = L_0(-r, t)$ for every s and t in S. Hence if f is in G_0 then there is an x in X such that $f = L_0(-r, t)x$. Thus we have reduced our problem to determining x at $t = 0$ from observations

$$[\eta L_0(-r, \ )x](s) = \begin{cases} \gamma L_0(-r, -r)x & -r \le s \le r' - r \\ \gamma L_0(-r, s - r')x & r' - r \le s \le 0 \end{cases}$$

Clearly, this task can be accomplished if there are d points $\{t_i\}^d$, in $[-r, -r']$ such that rank $[L_0(-r, t_1)^*\gamma^*, L_0(-r, t_2)^*\gamma^*, \cdots, L_0(-r, t_d)^*\gamma^*] = d$. For example, assume $r = \pi$, $r' = \pi/2$, $d = 2$, $L_0(-r, t) = \begin{bmatrix} \cos t & \sin t \\ -\sin t & \cos t \end{bmatrix}$ and $\gamma = \begin{bmatrix} 1 & 0 \\ 0 & 0 \end{bmatrix}$.

Then $\gamma L_0(-r, t) = \begin{bmatrix} \cos t & \sin t \\ 0 & 0 \end{bmatrix}$ and rank $\begin{bmatrix} -1 & 0 & 0 & 0 \\ 0 & 0 & -1 & 0 \end{bmatrix} = 2$,

i.e., if $f(t) = \begin{bmatrix} \cos t & \sin t \\ 0 & 0 \end{bmatrix} x$ then we can determine x from $[\eta f](-\pi)$ and $[\eta f](-\pi/2)$.

<u>Exercise</u>. If $r = r' = 0$ then the condition, i.e., the requirement that f can be determined at time $t = 0$ from observations ηf on $[-r, 0]$, must fail except in the trivial case when γ is nonsingular. Show that we can shift the origin in this case to get back to the previous situation, i.e., $r > 0$ and $r' = 0$.

The second problem, that of determining the implications for h given $[\eta h](t) = 0$ for $T - r + r' \le t \le T + r'$, can be brought into focus by the following result.

<u>Exercise</u>. Assuming f, h, and γ are as given in the optimal control problem and

solution, these are equivalent:

1. $h(t) = 0$ for $T - r \leq t \leq T$ and
2. $[AP_0 f](t)$ is in $cl(\mathbf{R}(\gamma^*))$ and $[\gamma h](t) = 0$ for $T - r \leq t \leq T$.

Summary

We have introduced finite interval optimal control problems for both basic and elaborated hereditary systems, i.e., both systems with and without input and observation processes. The solution methodology is based on an abstract formulation of the problems as constrained optimization problems in Hilbert spaces with use made of the reproducing kernel properties of the special spaces that provide the setting. The nonstandard cost functional, assigning control costs in terms of the system response for a system initially at rest with input the control, coupled with RKH space methods and the algebraic properties of the classes $\boldsymbol{A}$ and $\boldsymbol{B}$ yields explicit representations of the optimal feedback operators for this class of optimal control problems for hereditary systems.

Although the systems we consider can be given equivalent state space representations, we have not followed that path. Thus the complete theory available for finite dimensional state space systems, which serves as a guide for the semigroup/evolution equation approach [35], provides for us only goals and examples. In this sense, the material of this chapter is incomplete. Yet in another sense we do better.

Each of the following systems represents a step up in complexity for the semigroup/evolution equation approach.

1. $h(t) = f(t) + \int_0^t \alpha h(s)\, ds + \int_0^t \beta h(s - r)\, ds$

2. $h(t) = f(t) + \int_0^t \alpha(s)\, h(s)\, ds + \int_0^t \beta(s) h(s - r)\, ds$

3. $h(t) = f(t) + \int_0^t \alpha(t, s) h(s)\, ds + \int_0^t \beta(t, s) h(s - r)\, ds$

In order to support the algebraic properties of the classes $\boldsymbol{A}$ and $\boldsymbol{B}$ and, consequently, our vision of systems composed of networks of interacting components, we must be able to deal at once with all of these systems. That our results are complete in this sense provides the best justification of the path we have followed.

Variational methods have been applied to a range of optimal control problems for hereditary systems [3]. We identify the following as the principal difficulties encountered with this approach.

1. Dealing with point and terminal contraints.
2. Computing appropriate adjoints
3. Allowing for manipulation of the cost functional as part of some design methodology.

These problems are mitigated by the RKH space setting of Hellinger integrable functions and our choice of the form for the cost functional for hereditary systems.

In contrast, the semigroup or evolution equation formulation leads to an operator Riccati equation [10, 11, 19, 20] for a class of problems with some heredidtary dynamics but constraints of a restricted form. These results are satisfying in that they have a close analogy with a large body of finite dimensional state space results. However, since we want to carry forward the algebraic properties of the classes A and B and, in Chapter V, to allow for interval constraints, these results are not of particular interest here.

The sufficient conditions for weak controllability for systems with input processes are given in terms of the matrix representations of the input-response operators, i.e., in terms of operators and functions which are already part of the discussion. The conditions are "checkable" in the sense that they reduce the problem partially to that of testing a set of finite dimensional vector functions for linear independence. Weak controllability represents the minimum system requirement for our methods to work.

There is an extensive literature devoted to semigroup and evolution equation methods applied to this problem. The principal line of development for controllability and observability appears in [20, 27, 28, 29, 30, 31, 32]. These results are amplified with the required background filled in [36, 37]. Of course, the abstraction based on integral equations that we have adopted means that our controls are indefinite integrals of the controls in this literature.

The material of this chapter while of interest by itself was presented with an eye on what follows in Chapter V. Thus we focused on finite interval, terminal constraint, optimal control problems which we could solve explicitly. Here the emphasis is on central controller, but there we will adapt these methods to the decentralized case. The next chapter takes up the numerical problems associated with the methods introduced here. In Chapter IV, we extend most of our results to stochastic systems.

References

1. N. Aronzajn, Theory of reproducing kernels, Am. Math. Soc. Trans., 68(1950), 337-404.

2. H. T. Banks, Control of functional differential equations with function space boundary conditions, Delay and Functional Differential Equations and their Applications, K, Schmitt, ed., Academic Press, New York, 1972, 1-16.

3. H. T. Banks and A. Manitius, Application of abstract variational theory to hereditary systems - A survey, IEEE Trans. Auto. Cont., AC-19(1974), 524-533.

4. H. T. Banks, Function space controllability for linear functional differential equations, Proc. Conf. on Differential Games and Control Theory, Marcel Dekker, New York, 1974.

5. H. T. Banks, M. C. Jacobs, C. E. Langenhop, Characterization of the controlled states in $W_2^{(1)}$ of linear hereditary systems, SIAM J. Cont. 13(1975), 611-649.

6. Z. Bartosiewicz, Approximate controllability of neutral systems with delays in control, J. Diff. Eqs., 51(1984), 295-325.

7. Z. Bien and D. H. Chyung, Optimal control of delay systems with final function condition, Int. J. Cont., 32(1980), 539-560.

8. W. L. Chan and C. W. Li, Controllability, observability and duality in infinite dimensional linear systems with multiple norm-bounded controllers, Int. J. Cont., 33(1981), 1039-1058.

9. F. Colonius and D. Hinrichsen, Optimal control of functional differential systems, SIAM J. Cont. and Opt., 16(1978), 861-879.

10. R. F. Curtain, The infinite-dimensional Riccati equations with applications to affine hereditary differential systems, SIAM J. Cont., 13(1975), 1130-1143.

11. R. F. Curtain and A. J. Pritchard, Infinite Dimensional Linear Systems Theory, Springer-Verlag, Berlin and New York, 1978.

12. R. F. Curtain, Linear-quadratic control problem with fixed endpoints in infinite dimensions, J. Opt. Thy and Appl., 44(1984), 55-74.

13. M. C. Delfour and S. K. Mitter, Controllability, observability and optimal feedback control of hereditary differential systems, SIAM J. Cont., 10(1972), 298-328.

14. M. C. Delfour, C. McCalla, and S. K. Mitter, Stability of the infinite-time quadratic cost problem for linear differential systems, SIAM J. Cont., 13(1975), 48-88.

15. M. C. Delfour, The linear quadratic control problem for hereditary differential systems: Theory and numerical solution, Appl. Math. and Opt., 3(1977), 101-162.

16. M. C. Delfour, E. B. Lee, and A. Manitius, F-reduction of the operator Riccati equation for hereditary differential systems, Automatica, 14(1979), 385-395.

17. M. C. Delfour, Linear optimal control of systems with state and control variable delays, Automatica, 20(1984), 69-77.

18. S. Dolecki and D. L. Russell, A general theory of observation and control, SIAM J. Cont. and Opt., 15(1977), 185-220.

19. J. S. Gibson, The Riccati integral equations for optimal control problems on Hilbert spaces, SIAM J. Cont. and Opt., 17(1979), 537-565.

20. J. S. Gibson, Linear quadratic optimal control of hereditary differential systems - Infinite dimensional Riccati equations and numerical approximations, SIAM J. Cont. and Opt., 21(1983), 95-139.

21. G. Gripenburg, The construction of the solution of an optimal control problem described by a Volterra integral equation, SIAM J. Cont. and Opt., 21(1983), 582-597.

22. A. Ichikawa, Quadratic control of evolution equations with delays in control, SIAM J. Cont. and Opt., 20(1982), 645-668.

23. R. E. Kalman, P. L. Falb, and M. A. Arbib, Topics in mathematical system theory, McGraw-Hill, New York, 1969.

24. J. C. Louis and D. Wexler, On exact controllability in Hilbert spaces, J. Diff. Equations, 49(1983), 258-269.

25. D. G. Luenburger, Optimization by Vector Space Methods, John Wiley, New York, 1969.

26. A. Manitius and A. Olbrot, Controllability conditions for linear systems with delayed state and control, Arch. Automat. i. Telemech., 17(1972), 119-131.

27. A. Manitius and R. Triggiani, Function space controllability of linear retarded systems: A derivation from abstract operator conditions, SIAM J. Cont. and Opt., 16(1978), 599-645.

28. A. Manitius, Necessary and sufficient conditions of approximate controllability for general linear retarded systems, SIAM J. Cont. and Opt., 19(1981), 516-532.

29. A. Manitius, F-controllability and observability of linear retarded systems, Appl. Math. Opt., 9(1982), 73-95.

30. A. W. Olbrot, Control of retarded systems with function space constraints: Necessary optimality conditions, Control and Cybernetics, 5(1976), 5-31.

31. A. W. Olbrot, Control of retarded systems with function space constraints, Part 2: Approximate controllability, Control and Cybernetics 6(1977), 17-69.

32. A. W. Olbrot, Control to equilibrium of linear delay-differential systems, IEEE Trans. Auto. Cont., AC-28(1983), 521-523.

33. D. W. Ross and I. Flügge-Lotz, An optimal control problem for systems with differential-difference equation dynamics, SIAM J. Cont., 7(1969), 609-623.

34. D. W. Ross, Controller design for time lag systems via a quadratic criterion, IEEE Trans. Auto. Cont., AC-16(1971), 644-672.

35. D. L. Russell, Mathematics of Finite Dimensional Control Systems: Theory and Design, Marcel Dekker, New York, 1978.

36. D. Salamon, On controllability and observability of time delay systems, IEEE Trans. Auto. Cont., AC-29(1984), 432-439.

37. D. Salamon, Control and observation of neutral systems, Pitman Advanced Publishing Program, Boston, 1984.

38. K. Watanabe, M. Ito, and M. Kaneko, Finite spectrum assignment problem for systems with multiple commensurate delays in state variables, Int. J. Cont., 38(1983), 913-926.

39. K. Watanabe, Further study of spectral controllability of systems with multiple commensurate delays in state variables, Int. J. Cont., 39(1984), 497-505.

40. K. Watanabe, M. Ito, and M. Kaneko, Finite spectrum assignment problem of systems with multiple commensurate delays in states and controls, Int. J. Cont., 39(1984), 1073-1082.

41. R. B. Vintner and R. H. Kwong, The infinite time quadratic control problem for linear systems with state and control delays: An evolution equation approach, SIAM J. Cont. and Opt., 19(1981), 139-153.

III
Operator Approximation and Related Problems

1. Introduction

In this chapter, we consider the approximation of system operators and the reduction of several system problems to a setting suitable for numerical computations. The analysis takes place in a reproducing kernel Hilbert space of Hellinger integrable functions; see Chapter I, Section 4. Within this space a dense subset of polygonal functions arises naturally and standard projection methods can be used to obtain finite dimensional approximations to system operators and to solutions of system equations.

Convergence of approximations to operators in the classes $\mathcal{A}$ and $\mathcal{B}$ is considered. Since operators in $\mathcal{A}$ need not be compact, i.e., the identity operator belongs to $\mathcal{A}$, convergence cannot be uniform. Consequently, an alternate operator norm is introduced. Examples illustrating the discrete structure of operator approximations, computational procedures, and rates of convergence are included. Ideas arising in the approximation of operators are further developed in the analysis of identification and parameter estimation problems and in the numerical solution of optimal control problems.

The system identification problem is to determine the matrix representation of a system operator based upon a limited amount of information. The case of linear, time invariant operators is considered and an algorithm is developed for determining such an operator from knowledge of input/output (input/response) data. Throughout this chapter, if h = Af with A in $\mathcal{A}$ then the pair (f, h) is referred to as an <u>input/output</u> or <u>input/response</u> pair. The parameter estimation problem is to determine approximations to parameters appearing in an operator equation description of the system using input-output data. Many computational simplifications are illustrated in the parameter estimation problem for delay differential equations. In the analysis of identification and estimation problems, it is assumed that input/output data is known exactly, i.e., stochastic considerations are not taken into account at this point. The chapter concludes with a consideration of numerical solutions to the optimal control problems studied in Chapter II.

Throughout this chapter, let $S = [-r, \infty)$ and X denote the space of real d-tuples with the Euclidean norm $|\cdot|$ arising from the usual inner product $<\cdot, \cdot>$. Elements of X will be represented as column vectors, coordinates with respect to the standard basis $\{e_1, ..., e_d\}$ of X, and for x and y in X, $<x, y> = x^*y$ where the asterisk denotes transpose. Let I_d denote the identity transformation on X.

The underlying function space G will consist of those X valued functions f defined on S which are continuous on $[-r, 0)$ and $[0,\infty)$ and for which f(0-) exists. For f in G and v in S^+, define

$$[P_v f](u) = \begin{cases} f(u) & -r \leq u \leq v \\ f(v) & v \leq u \end{cases}$$

Also, let $N_v(f) = \sup_{-r \leq t \leq v} | f(t) |$. It follows that $\{G, N\}$ is complete and, consequently, $\{S, G, N, P\}$ satisfies the underlying function space properties for a linear hereditary system. See Chapter I.

In this chapter we want to assume in each discussion that k is a fixed nondecreasing function of the form $k(t) = 1 + r + r_1 + t$ for $t \geq 0$, where $r_1 \geq 0$. This allows us to speak of the space G_H of functions from S into X which are Hellinger integrable with respect to k without introducing any special notation indicating the dependence of G_H on k. Furthermore, we will modify condition 2) in the definition of $\boldsymbol{B}$, see Chapter I, as follows:

ii') for each subinterval [u, v] of $[0,\infty)$ there is a number b such that $| [Bf](t) - [Bf](s) | \leq b\int_{[s,t]} N_x(f)\, dk(x)$ for each f in G and subinterval [s, t] of [u, v].

The nondecreasing function bk is called a <u>variation</u> <u>for</u> $\boldsymbol{B}$ <u>on</u> [u, v].

With this modification in the definitions of $\boldsymbol{A}$ and $\boldsymbol{B}$ the basic properties of elements of $\boldsymbol{A} \cup \boldsymbol{B}$ and the basic relationship between $\boldsymbol{A}$ and $\boldsymbol{B}$ still hold, Theorems 1.1 and 1.2. One should note for each T in S that $P_T G_H \subset G_\infty \subset G_H \subset G$, $N_T(f) \leq$

$N_H(P_Tf)k(T)^{1/2}$ for all f in G_H, and if k(0) - k(0-) > 0 then P_TG_H is dense in P_TG with respect to the sup norm N_T. Also each B in ***B*** maps G into G_H and for f in G_H there is a unique h in G_H such that h = f + Bh. Hence $A = (I - B)^{-1}$ maps G_H onto G_H. For T in S both P_TA and P_TB are continuous transformations of P_TG_H with respect to the norms N_H and N_T, see Theorem 1.22. Recall that the reproducing kernel for $\{G_\infty, Q_H\}$ is defined by K(u, v) = k(min(u, v)) for $-r \le u, v$.

In this chapter, we are primarily concerned with finite interval problems, i.e., functions h defined on S which satisfy h(t) = f(t) + [Bh](t) for $-r \le t \le T$ for a given f in P_TG_H. Of course if h satisfies h = f + Bh on S then h satisfies the finite interval problem for each subinterval [-r, T] of S. If f is Hellinger integrable on [-r, T] and is extended to S by defining f(t) = f(T) for $T \le t$ then f is in G_∞ and the solution of the equation h = f + Bh restricted to [-r, T] solves the finite interval problem. Clearly the finite interval problem is equivalent to solving equations of the form $h = P_Tf + P_TBh$ or $h = P_TAP_Tf$. Since $P_TG_H \subset G_\infty$ the space $\{G_\infty, N_H\}$ provides a unified setting for the analysis of approximation and discretization problems.

As in Section 4 of Chapter I, operators F in ***A*** or ***B*** have a matrix representation LF such that for each T in S^+, x in X, f in G_H and t in [-r, T]

$$<[Ff](t), x> = <[P_TFf](t), x> = Q_H(f, LF(, t)x) \tag{3.1.1}$$

where $LF(v, u)x = [(P_TF)^*K(, u)x](v)$ for $-r \le u, v \le T$. To obtain LF without having to compute $(P_TF)^*$ note that

$$\begin{aligned} LF(u, v)_{ij} &= <LF(u, v)e_j, e_i> \\ &= Q_H(LF(, v)e_j, K(, u)e_i) \\ &= Q_H((P_TF)^*K(, v)e_j, K(, u)e_i) \\ &= Q_H(K(, v)e_j, P_TFK(, u)e_i) \\ &= <[FK(, u)e_i](v), e_j> \end{aligned} \tag{3.1.2}$$

where $\{e_1, e_2, ..., e_d\}$ denotes the standard basis for X .

We make no distinction in notation between LF(u, v) as a linear transformation on X and its matrix representation with respect to the basis $\{e_1, \cdots, e_d\}$. From (3.1.2) one obtains for $-r \le u, v$

$$LF(u, v)^* = [FK(\, , u)](v) \tag{3.1.3}$$

where $[FK(\, , u)](v)$ is the d×d matrix with j, i-th component $<[FK(\, , u)e_i](v), e_j>$. In light of this last equation we speak of the matrix representation of F rather than $P_T F$.

From a systems viewpoint it is most important to observe from equations (3.1) that operators in $\boldsymbol{A}$ or $\boldsymbol{B}$ are completely determined by their response to the inputs $K(\, , u)e_i$ for u in S and i = 1, •••, d. This observation plays a central role in the development of operator approximations and in the analysis of identification and parameter estimation problems to follow.

The causality of operators F in $\boldsymbol{A}$ or $\boldsymbol{B}$ implies $<[Ff](t), x> = <[FP_t f](t), x>$ for each t in [-r, T] and x in X. Consequently,

$$\begin{aligned} <[Ff](t), x> &= Q_H(P_t f, LF(\, , t)x) \\ &= <f(-r), LF(-r, t)x> + \int_{-r}^{t} <df, dLF(\, , t)x>/dk \end{aligned} \tag{3.2}$$

or using vector matrix notation

$$[Ff](t) = LF(-r, t)^* f(-r) + \int_{-r}^{t} dLF(\, , t)^* df/dk$$

Other properties of the matrix representation of operators A in $\boldsymbol{A}$ and B in $\boldsymbol{B}$ to keep in mind throughout this chapter are

$$LA(u, v) = K(v, u), \;\; LB(u, v) = 0I_d \qquad v \leq 0 \tag{3.3}$$

$$LA(u, v) = LA(v, v), \;\; LB(u, v) = LB(v, v) \qquad 0 \leq v \leq u \tag{3.4}$$

If A and B are time invariant, see Chapter I, then

$$\begin{aligned} LA(v, w) - LA(u, w) &= LA(v + b, w + b) - LA(u + b, w + b) \\ LB(v, w) - LB(u, w) &= LB(v + b, w + b) - LB(u + b, w + b) \end{aligned} \tag{3.5}$$

for $0 \leq u \leq v \leq w$ and $b \geq 0$.

Example 3.1. To illustrate the matrix representation of some simple operators, consider the system

$$h(t) = \begin{cases} f(t) & -r \le t \le 0 \\ f(t) + \int_0^t \alpha h(\tau - \rho)d\tau & 0 \le t \end{cases}$$

with $0 \le \rho \le r$ and α a $d \times d$ matrix. Here

$$[Bh](t) = \begin{cases} 0 & -r \le t \le 0 \\ \int_0^t \alpha h(\tau - \rho)\, d\tau & 0 \le t \end{cases}$$

For this example $k(t) = 1 + r + t$ is a variation for B. The matrix representation of B is

$$LB(u, v) = \begin{cases} 0 & -r \le u & -r \le v \le 0 \\ \alpha^*(1 + r + u)v & -r \le u \le -\rho & 0 \le v \\ \alpha^*[(1 + r + u)v - (u + p)^2/2] & -\rho \le u \le v - \rho & 0 \le v \\ \alpha^*[(1 + r - \rho)v + v^2/2] & v - \rho \le u & 0 \le v \end{cases} \tag{3.6}$$

The input-output operator A for this example is the set of ordered pairs to which (f, h) belongs only in case

$$h(t) = \begin{cases} f(t) & -r \le t \le 0 \\ f(t) + \sum_{n=0}^{m} \int_{n\rho}^{t} ((t - s)^n/n!)\alpha^{n+1} f(s - (n + 1)\rho)\, ds & m\rho \le t \le (m + 1)\rho \end{cases}$$

For $-r \le v \le 2\rho$ one obtains

$$LA(u, v) = \begin{cases} K(v, u) & -r \le v \le 0 \\ K(v, u) + \int_0^v \alpha K(s - \rho, u)\, ds & 0 < v \le \rho \\ K(v, u) + \int_0^v \alpha K(s - \rho, u)\, ds + \int_\rho^v (v - s)\alpha^2 K(s - 2\rho, u)\, ds & \rho < v \le 2\rho \end{cases}$$

Exercise. For this example verify that $S_{-b}BS_bf = Bf$ and $S_{-b}AS_bf = Af$ for f in $(I - P_{-\rho})G_H$ and $b \ge 0$, where the shift operators S_b and S_{-b} are as defined in Section

4 of Chapter I. It follows that LA(v, w) - LA(u, w) = LA(v + b, w + b) - LA(u + b, w + b) whenever $-\rho \leq s \leq v \leq w$. The same identity holds for LB.

Exercise. Show that $k(t) = 1 + r + t$ for $-r \leq t < 0$ and $k(t) = 1 + 2r + t$ for $0 \leq t$ is a variation for B as defined in Example 3.1.

Exercise. Let $k(t) = 1 + r + t$ for $-r \leq t$, α be a d×d matrix and $[Bh](t) = 0$ for $-r \leq t \leq 0$.

a) If $[Bh](t) = \alpha\int_{[0,t]} h(\tau)\, d\tau$ for $0 \leq t$ verify that

$$LB(0, t) = \alpha^*(1 + r)t \text{ for } 0 \leq t \text{ and}$$

$$LB(c, t) = \begin{cases} \alpha^*[(1 + r)t + t^2/2] & 0 \leq t \leq c \\ \alpha^*[(1 + r + c)t - c^2/2] & c \leq t \end{cases}$$

b) If $[Bh](t) = \alpha\int_{[0,t]} (t - \tau)h(\tau)\, d\tau$ for $0 \leq t$ verify that

$$LB(0, t) = \alpha^*(1 + r)t^2/2 \text{ for } 0 \leq t \text{ and}$$

$$LB(c, t) = \begin{cases} \alpha^*[(1 + r)t^2/2 + t^3/6] & 0 \leq t \leq c \\ \alpha^*[(1 + r)t^2/2 + c^3/6 + ct^2/2 - tc^2/2] & c \leq t \end{cases}$$

c) If $[Bh](t) = \alpha\int_{[0,t]} h(\tau - \rho)\, d\tau$ for $0 \leq t$ verify that

$$LB(0, t) = \begin{cases} \alpha^*[(1 + r - \rho)t + t^2/2] & 0 \leq t \leq \rho \\ \alpha^*[(1 + r)t - \rho^2/2] & \rho \leq t \end{cases}$$

and

$$LB(c, t) = \begin{cases} \alpha^*[(1 + r - \rho)t + t^2/2] & 0 \leq t \leq \rho + c \\ \alpha^*[(1 + r + c)t - (c + \rho)^2/2 & \rho + c \leq t \end{cases}$$

2. K-Polygonal Functions

In this section, we explore further properties of the class of K-polygonal functions, which were introduced in Section 4 of Chapter II. Such polygonal functions arise naturally in reproducing kernel Hilbert spaces and are used along with projection methods to develop approximations to system operators.

Recall any function f on S of the form $f(t) = \Sigma_{p=0,n}K(t, s_p)x_p$ where $s = \{s_p\}_{p=0,n}$ is an increasing sequence in S and $\{x_p\}_{p=0,n}$ is a sequence of vectors in X is called a K-polygonal function. Clearly, any K-polygonal function is in G_∞ since each component is a linear combination of functions in G_∞. Also the subspace of all

K-polygonal functions determined by a fixed partition s of [-r, T] is a closed linear subspace of G_∞. As before we let Π_s denote the orthogonal projection of G_∞ onto this subspace. The union of the finite dimensional subspaces $\Pi_s G_\infty$ is dense in $P_T G_H$ with respect to the Hellinger norm N_H. To see why this density property holds, suppose f is in $P_T G_H$ and is orthogonal to K(, t)x for each t in [-r, T] and x in X. Then $\langle f(t), x\rangle = Q_H(f, K(, t)x) = 0$ and f is the zero function. The orthogonal complement of the closure of the union of the subspaces $\Pi_s G_\infty$ contains only the zero function. It follows that this union is dense in $P_T G_H$. One should also note for f in G_∞ and x in X that $\langle [\Pi_s f](s_p), x\rangle = Q_H(\Pi_s f, K(, s_p)x) = Q_H(f, K(, s_p)x) = \langle f(s_p), x\rangle$. Hence $\Pi_s f$ interpolates f at s_p in s.

If $f = \Sigma_{p=0,n} K(, s_p)x_p$ is a K-polygon and $s_{p-1} \le t \le s_p$ then

$$
\begin{aligned}
&\{k(t) - k(s_{p-1})\}f(s_p) + \{k(s_p) - k(t)\}f(s_{p-1}) \\
&\quad = \{k(t) - k(s_{p-1})\}\{k(s_0)x_0 + \dots + k(s_{p-1})x_{p-1} + k(s_p)x_p + \dots + k(s_p)x_n\} \\
&\qquad + \{k(s_p) - k(t)\}\{k(s_0)x_0 + \dots + k(s_{p-1})x_{p-1} + k(s_{p-1})x_p + \dots + k(s_{p-1})x_n\} \\
&\quad = \{k(s_p) - k(s_{p-1})\}\{k(s_0)x_0 + \dots + k(s_{p-1})x_{p-1}\} \\
&\qquad + \{k(s_p) - k(s_{p-1})\}\{k(t)x_p + \dots + k(t)x_n\} \\
&\quad = \{k(s_p) - k(s_{p-1})\}f(t)
\end{aligned}
$$

Hence any K-polygonal function may be represented piecewise by

$$
f(t) = \begin{cases} k(t)f(-r) & -r \le t \le s_0 \\ \{dk(s_{p-1}, t)f(s_p) + dk(t, s_p)f(s_{p-1})\}/dk(s_{p-1}, s_p) & s_{p-1} \le t \le s_p \\ f(s_n) & s_n \le t \end{cases} \tag{3.6}
$$

Moreover given a function f from S to X and a partition $s = \{s_p\}_{p=0,n}$ of [-r, T] there exists a unique K-polygonal function which interpolates f at each s_p in s. To see this consider the linear system of equations $f(s_p) = \Sigma_{q=0,n} K(s_p, s_q)x_q$ for $0 \le p \le n$. Let K_s denote the $(n + 1)d \times (n + 1)d$ block matrix whose (p, q) block is given by $K_s(p, q)_{i,j} = \langle e_i, K(s_p, s_q)e_j\rangle$ for $0 \le p,\ q \le n$ and $1 \le i,\ j \le d$. Let $f_s = (f(s_0)^*, f(s_1)^*, \cdots, f(s_n)^*)^*$ and $x = (x_0^*, x_1^*, \cdots, x_n^*)^*$. The matrix K_s has the following form

$$
\begin{bmatrix}
k(s_0) & k(s_0) & k(s_0) & \cdot & \cdot & \cdot & \cdot & k(s_0) \\
k(s_0) & k(s_1) & k(s_1) & \cdot & \cdot & \cdot & \cdot & k(s_1) \\
k(s_0) & k(s_1) & k(s_2) & \cdot & \cdot & \cdot & \cdot & k(s_2) \\
\cdot & \cdot & \cdot & \cdot & & & & \\
\cdot & \cdot & \cdot & & \cdot & & & \\
\cdot & \cdot & \cdot & & & \cdot & & \\
\cdot & \cdot & \cdot & & & & \cdot & \\
k(s_0) & k(s_1) & k(s_2) & & & & & k(s_n)
\end{bmatrix} \otimes I_d
$$

where $\otimes$ denotes the Kronecker product of the matrices and I_d the $d \times d$ identity matrix. With this notation the above system of equations becomes $f_s = K_s x$.

$$
\begin{bmatrix}
1+\frac{1}{dk(s_0, s_1)} & -\frac{1}{dk(s_0, s_1)} & 0 & \cdot & \cdot & \cdot & \cdot \\
-\frac{1}{dk(s_0, s_1)} & -\frac{1}{dk(s_0, s_1)}+\frac{1}{dk(s_1, s_2)} & -\frac{1}{dk(s_1, s_2)} & 0 & & & \\
0 & -\frac{1}{dk(s_1, s_2)} & \frac{1}{dk(s_1, s_2)}+\frac{1}{dk(s_2, s_3)} & -\frac{1}{dk(s_2, s_3)} & 0 & \cdot & \cdot \\
 & 0 & -\frac{1}{dk(s_2, s_3)} & \cdot & \cdot & & \\
 & & & & \cdot & \cdot & \\
 & & & \cdot & \cdot & \cdot & -\frac{1}{dk(s_{n-1}, s_n)} \\
 & & & & -\frac{1}{dk(s_{n-1}, s_n)} & & +\frac{1}{dk(s_{n-1}, s_n)}
\end{bmatrix} \otimes I_d
$$

One can show that $(K_s)^{-1}$ is a symmetric, block tridiagonal matrix. For $p = 1, \cdots, n$ the p-th block of the subdiagonal is $-(1/dk(s_{p-1}, s_p))I_d$. The first main diagonal block is $(1/k(s_0) + 1/dk(s_0, s_1))I_d$, the last main diagonal block is $(1/dk(s_{n-1}, s_n))I_d$, and for $p = 2, \cdots, n$ the p-th main diagonal block is $\{1/dk(s_{p-2}, s_{p-1}) + 1/dk(s_{p-1}, s_p)\}I_d$. That is $(K_s)^{-1}$ has the form above.

The matrix $(K_s)^{-1}$ can be simplified when k is restricted as in Section 1 and the partition s of [-r, T] is chosen so that $s_p = -r + pc$ for p = 0, 1, ..., n with r = p'c for some positive integer p'. In this case, $(K_s)^{-1}$ is the block tridiagonal matrix.

$$\frac{1}{c}\begin{bmatrix} c+1 & -1 & & & & & & & & \\ -1 & 2 & -1 & & & & & & & \\ & \bullet & \bullet & \bullet & & & & & & \\ & & -1 & 2 & -1 & & & & & \\ & & & -1 & 1+\frac{c}{r_1+c} & -\frac{c}{r_1+c} & & & & \\ & & & & -\frac{c}{r_1+c} & \frac{c}{r_1+c}+1 & -1 & & & \\ & & & & & -1 & 2 & -1 & & \\ & & & & & & \bullet & \bullet & \bullet & \\ & & & & & & & -1 & 2 & -1 \\ & & & & & & & & -1 & 1 \end{bmatrix} \otimes I_d$$

Thus $f_s = K_s x$ may be solved uniquely for x and there is a unique K-polygon which interpolates f at each s_p in s. For f in G_H we let $\Pi_s f$ denotes the K-polygonal function which interpolates f at each s_p in s; i.e., Π_s is extended to G_H.

For $s = \{s_p\}_{p=0,n}$ a partition of [-r, T] we let θ_s denote the mapping from $R^{d(n+1)}$ to $\Pi_s G_H$ which maps x in $R^{d(n+1)}$ to the K-polygonal function f which satisfies $f(s_p) = x_p$ for each s_p in s. Thus for f in G_H we have $\theta_s f_s = \Pi_s f$.

Remark. For f in G_H the existence of a unique-polygonal function which interpolates f at each s_p in s follows from the Projection Theorem. If g is the element of $\Pi_s G_H$ closest to $P_T f$ then necessarily $Q_H(P_T f - g, K(, s_q)x) = 0$ for each s_q in s and x in X. From the reproducing kernel property we obtain $f(s_q) = [P_T f](s_q) = g(s_q)$ and consequently $g = \Pi_s f$.

It should be noted that the inner product Q_H can be evaluated as a finite sum when one of its arguments is from $\Pi_s G_H$. This property is the basis for the numerical evaluation of approximations to be introduced.

Lemma 3.1. For f in $\Pi_s G_H$ and g in G_H,

$$Q_H(f, g) = <f(-r), g(-r)> + \Sigma_s <df, dg>/dk \tag{3.7}$$

Proof. Let $t = \{t_q\}$ be a refinement of a partition $s = \{s_q\}$ of $[-r, T]$. Then for $s_{p-1} \le t_{q-1} < t_q \le s_p$ with s_{p-1}, s_p in s and t_{q-1}, t_q in t it follows from (3.5) that $f(t_q) - f(t_{q-1}) = dk(t_{q-1}, t_q)df(s_{p-1}, s_p)/dk(s_{p-1}, s_p)$ and a term of the approximating sum $\Sigma_t <df, dg>/dk$ for $Q_H(f, g)$ becomes $<df(s_{p-1}, s_p), dg(t_{q-1}, t_q)>/dk(s_{p-1}, s_p)$. Summing for all q such that $s_{p-1} \le t_{q-1} < t_q \le s_p$ yields $<df(s_{p-1}, s_p), dg(s_{p-1}, s_p)> / dk(s_{p-1} s_p)$. Since t is an arbitrary refinement of s the result follows. □

As a consequence of this Lemma note that, for f in G_H, $N_H(P_T f - \Pi_s f) = Q_H(P_T f, P_T f) - <f(-r), f(-r)> - \Sigma_s <df, df>/dk$. Hence $\Pi_s f$ converges to $P_T f$ in $\{G_\infty, N_H\}$ with respect to refinements of partitions of $[-r, T]$.

One may also rewrite equations (3.7) in a matrix form suitable for numerical computations. For f and g in G_H and s a partition of $[-r, T]$, the Hellinger inner product of $\Pi_s f$ and g can be written as

$$Q_H(\Pi_s f, g) = f_s^*(K_s)^{-1} g_s \tag{3.8}$$

The derivation of (3.8) requires an expansion of equation (3.7) and regrouping of terms as follows:

$$Q_H(\Pi_s f, g) = <f(s_0, g(s_0)> + \sum_{p=1}^{n} <df(s_{p-1}, s_p), dg(s_{p-1}, s_p)>/dk(s_{p-1}, s_p)$$

$$= <f(s_0), g(s_0)> + \sum_{p=1}^{n} <f(s_p), dg(s_{p-1}, s_p)>/dk(s_{p-1}, s_p)$$

$$- \sum_{p=1}^{n} <f(s_{p-1}), dg(s_{p-1}, s_p)>/dk(s_{p-1}, s_p)$$

$$= <f(s_0), g(s_0)> + <f(s_n), dg(s_{n-1}, s_n)>/dk(s_{n-1}, s_n)$$

$$+ \sum_{p=1}^{n-1} <f(s_p), dg(s_{p-1}, s_p)/dk(s_{p-1}, s_p)$$

$$- \langle f(s_0), dg(s_0, s_1)\rangle/dk(s_0, s_1) - \sum_{p=1}^{n} \langle f(s_{p-1}), dg(s_{p-1}, s_p)\rangle/dk(s_{p-1}, s_p)$$

$$= \langle f(s_0), g(s_0) - dg(s_0, s_1)/dk(s_0, s_1)\rangle$$

$$+ \sum_{p=1}^{n-1} \langle f(s_p), - g(s_{p-1})/dk(s_{p-1}, s_p)$$

$$+ \{1/dk(s_{p-1}, s_p) + 1/dk(s_p, s_{p+1})\}g(s_p) - g(s_{p+1})/dk(s_p, s_{p+1})\rangle$$

$$+ \langle f(s_n), dg(s_{n-1}, s_n)\rangle/dk(s_{n-1}, s_n)$$

$$= f_s^*(K_s)^{-1}g_s$$

Since $(K_s)^{-1}$ is tridiagonal, the matrix multiplication necessary for evaluation of $Q_H(f_s, g) = f_s^*(K_s)^{-1}g_s$ can be greatly simplified. Further numerical consequences will become apparent.

3. Operator Approximations

An approximation of operators in ***A*** and ***B*** which reduces the analysis of many system problems to a finite dimensional setting is now considered. The approach is based upon projection methods [19, 21, 24]. Our reproducing kernel Hilbert space setting provides a structure for the calculation of such approximations for a broad class of linear hereditary systems.

For F in ***A*** or ***B*** and s a partition of [-r, T] the operator $\Pi_s F \Pi_s$ may be used to approximate F. It should be noted that $[\Pi_s F \Pi_s f](s_p) = [\Pi_s Ff](s_p) = [Ff](s_p)$ whenever f is in $\Pi_s G_H$ and s_p is in s. Thus for f in $\Pi_s G_H$ both $\Pi_s F \Pi_s f$ and Ff take on the same values at partition points. Also, if M is the matrix representation of $\Pi_s F \Pi_s$ and f is in $\Pi_s G_H$, then, using (3.8),

$$\langle [\Pi_s F \Pi_s f](s_p), x\rangle = Q_H(f, M(\, , s_p)x)$$
$$= \langle f(-r), M(-r, s_p)x\rangle + \sum_s \langle df, dM(\, , s_p)x\rangle/dk \qquad (3.9)$$

or using matrix notation

$$[\Pi_s F \Pi_s f](s_p) = M(-r, s_p)^*f(-r) + \sum_s dM(\, , s_p)^*df/dk$$

This equation is the discrete analog to equation (3.2). Moreover, $M(s_p, s_q) = LF(s_p, s_q)$ whenever s_p and s_q are in s.

Exercise. Verify that equation (3.9) can be written as

$$<[\Pi_s F \Pi_s f](s_p), x> = <f(s_0), LF(s_0, s_p)x - (LF(s_1, s_p)x - LF(s_0, s_p)x)/dk_1>$$

$$+ \sum_{q=1}^{p-1} <f(s_q), LF(s_q, s_p)x - LF(s_{q-1}, s_p)x)/dk_q - LF(s_{q+1}, s_p)x - LF(s_q, s_p)x/dk_{q+1}>$$

with s_p in s.

Exercise. If F in (3.9) is time invariant, f is in $(I - P_0)G_H$, and $s_p = pc$ for $p = 0, 1, \cdots, n$ with $r = p'c$ and $T = nc$ then

$$<[\Pi_s F \Pi_s f](c), x> = <f(c), (LF(c, c)x - LF(0, c)x)/dk_1>$$

and for $p = 2, \cdots, n$

$$<\Pi_s F \Pi_s f](pc), x> = \sum_{q=1}^{p-1} <f(qc), (LF(c, (p - q + 1)c)x - LF(0, (p - q + 1)c)x)/dk(q-1)c, qc)$$

$$- (LF(c, (p - q)c)x - LF(0, (p - q)c)x)/dk(qc, (q + 1)c)$$

$$+ <f(pc), (LF(c, c)x - LF(0, c)x)/dk(0, c)>$$

Consequently the time invariance assumption can lead to a reduction in storage requirements necessary for calculation of $\Pi_s F \Pi_s f$.

The convergence of the approximations $\Pi_s F \Pi_s$ to $P_T F$ with F in ***A*** or ***B*** is now considered. Since $\Pi_s F \Pi_s$ is compact and $P_T F$ might not be compact, our approximation cannot be in the sense of the operator norm N_H. Let N_{HT} denote the operator pseudonorm defined on the linear operators F of G_H which are continuous with respect to N_T by

$$N_{HT}(F) = \sup\{N_T(Ff)/N_H(P_T f): \text{ f in } G_H \text{ and } P_T f \neq 0\}$$

Theorem 3.2. The operator P_TF is the limit through refinement of the net of compact operators $\{\Pi_s F \Pi_s\}$, s a partition of [-r, T], with respect to the operator pseudo-norm N_{HT}.

We will show for each positive number ε there is a partition t′ of [-r, T] such that if t refines t′ then $| \langle [P_TFf](u) - [\Pi_t F \Pi_t f](u), x \rangle | \leq N_H(f) \, |x| \, \varepsilon$ for each u in [-r, T], x in X, and f in G_H. Letting $x = [P_TF](u) - [\Pi_t A \Pi_t f](u)$ the Theorem follows.

Let L denote the matrix representation of F.

Lemma 3.3. If t is a partition of [-r, T], M is the matrix representation of $\Pi_t F \Pi_t$, u is in the range of t, and x is in X then $N_H(L(\,,u)x - M(\,,u)x)^2 - |L(-r,u)x|^2 = N_H(L(\,,u)x)^2 - \Sigma_t |dL(\,,u)x|^2/dk$.

Proof. Since $M(\,,u)x = (\Pi_t F \Pi_t)^* K(\,,u)x = \Pi_t F^* \Pi_t K(\,,u)x = \Pi_t L(\,,u)x$ is in $\Pi_t G_H$ and $L(t_p, t_q) = M(t_p, t_q)$ for t_p, t_q in the range of t, then using Lemma 3.1 it follows that $Q_H(L(\,,u)x, M(\,,u))x) = |L(-r,u)x|^2 + \Sigma_t |dL(\,,u)x|^2/dk$. Thus

$$\begin{aligned} N_H(L(\,,u)x - M(\,,u)x)^2 &= Q_H(L(\,,u)x - M(\,,u)x, L(\,,u)x - M(\,,u)x) \\ &= N_H(L(\,,u)x)^2 - |L(-r,u)x|^2 - \Sigma_t |dL(\,,u)x|^2 dk \quad \square \end{aligned}$$

Lemma 3.4. If [u, v] is a subinterval of [-r, T] and x is in X then $N_H(L(\,,v)x - L(\,,u)x) \leq N_H(P_TF) \, |x| \, dk(u,v)^{1/2}$. Furthermore, if t is a partition of [-r, T] and M represents $\Pi_t F \Pi_t$ then $N_H(M(\,,v)x - M(\,,u)x) \leq N_H(P_TF) \, |x| \, dk(u,v)^{1/2}$

Proof. From the representation theorem $N_H(L(\,,v)x - L(\,,u)x) = N_H((P_TF)^*\{K(\,,v)x - K(\,,u)x\}) \leq N_H(P_TF) \, |x| \, dk(u,v)^{1/2}$ and

$$\begin{aligned} N_H(M(\,,v)x - M(\,,u)x) &= N_H((\Pi_t F \Pi_t)^*\{K(\,,v)x - K(\,,u)x\}) \\ &\leq N_H(\Pi_t F \Pi_t) \, |x| \, dk(u,v)^{1/2} \\ &\leq N_H(P_TF) \, |x| \, dk(u,v)^{1/2} \quad \square \end{aligned}$$

Before proving Theorem 3.2, note that for each positive number ε there is a partition $s = \{s_p\}_{p=0,n}$ of [-r, T] such that if u lies in a subinterval $[s_{p-1}, s_p]$ then either $dk(u, s_p)^{1/2} < \varepsilon$ or $dk(s_{p-1}, u)^{1/2} < \varepsilon$.

Proof of Theorem 3.2. If $s = \{s_p\}_{p=0,n}$ is a partition of $[-r, T]$ with refinement t, M represents $\Pi_t F \Pi_t$, u in $[-r, T]$ satisfies $s_{p-1} \le u \le s_p$, and x is in X then

$$| <[P_T Ff](u) - [\Pi_t F \Pi_t f](u), x> |$$
$$= | Q_H(P_T f, L(, u)x - M(, u)x) |$$
$$\le N_H(P_T f)\{N_H(L(, u)x - L(, s_{p-1})x) + N_H(L(, s_{p-1})x - M(, s_{p-1})x)$$
$$+ N_H(M(, s_{p-1})x - M(, u)x)\}$$

For a given positive number ε, let s be a partition of $[-r, T]$ satisfying the conditions mentioned in the comments preceding this proof. Let x be in X and u in $[-r, T]$ satisfying $s_{p-1} \le u \le s_p$ and $dk(s_{p-1}, u)^{1/2} < \varepsilon$. A similar argument applies if we must take $dk(u, s_p)^{1/2} < \varepsilon$.

By Lemma 3.3, if $t = \{t_p\}_{p=0,m}$ refines s then

$$N_H(L(, s_p)x - M(, s_p)x)$$
$$\le | x | \{\sum_{i=1}^{d} N_H(L(, s_p)e_i - M(, s_p)e_i)^2\}^{1/2}$$
$$= | x | \{\sum_{i=1}^{d} (N_H(L(, s_p)e_i)^2 - | L(-r, s_p)e_i |^2 - \Sigma_t | dL(, s_p)e_i) |^2/dk)\}^{1/2}$$

Thus given $\varepsilon > 0$ there is a refinement t' of s such that if $t = \{t_p\}_{0,m}$ refines t', M represents $\Pi_t F \Pi_t$, and $p = 0, 1, \cdots, n$ then $N_H(L(, s_p)x - M(, s_p)x) \le | x | \varepsilon$. Note that t' depends on ε and s but not x.

Furthermore, from the above inequalities and Lemma 3.4 we obtain

$$| <[P_T Ff](u) - [\Pi_t F \Pi_t f](u), x> |$$
$$\le N_H(P_T f) \{N_H(P_T F) | x | dk(s_{p-1}, u)^{1/2} + | x | \varepsilon + N_H(P_T F) | x | dk(s_{p-1}, u)^{1/2}\}$$
$$\le N_H(P_T f) | x | \{N_H(P_T F) \varepsilon + \varepsilon + N_H(P_T F)\varepsilon\}$$

The result follows. □

Corollary 3.5. If A is in $\boldsymbol{A}$ and f is in G_H then Af is the limit through refinement of the net of K-polygonal functions $\{\Pi_s A \Pi_s f\}$ with respect to the norm N_T.

This Corollary follows since $N_T(Af - \Pi_s A \Pi_s f) \leq N_{HT}(A - A_s) N_H(P_T f)$.

If B is in $\boldsymbol{B}$, f is in G_H, and $s = \{s_p\}_{p=0,n}$ is a partition of [-r, T] then it is natural to compare solutions of the equation $h = f + Bh$ with solutions of $h = \Pi_s f + \Pi_s B \Pi_s h$ over the interval [-r, T]. Here $[\Pi_s B \Pi_s f](t) = \{(k(s_p) - k(t)) \cdot [B\Pi_s f](s_{p-1}) + (k(t) - k(s_{p-1}))[B\Pi_s f](s_p)\}/dk(s_{p-1}, s_p)$ when $s_{p-1} < t < s_p$ and so $\Pi_s B \Pi_s$ is not a causal operator. Consequently, the theory developed in Chapter I does not immediately guarantee the existence of a solution to the equation $h = \Pi_s f + \Pi_s B \Pi_s h$. In order to circumvent this problem we develop an alternate approximation.

Suppose that F is an element of $\boldsymbol{B}$, $\{t_p\}_{p=0,n}$ is a partition of [-r, T], and $c = \text{mesh}(t) = \max(t_p - t_{p-1})$, $p = 1, \cdots, n$. When $T > 0$ we assume that 0 is in the par- tition. Let F_t denote the linear transformation of G defined by

$$[F_t f](u) = \begin{cases} 0 & -r \leq u \leq c - r \\ [\Pi_t F \Pi_t f](u - c) & c - r \leq u \end{cases} \tag{3.10}$$

for each f in G. It will be shown that solutions of $h = \Pi_t f + F_t h$ can be used to approximate solutions of $h = f + Fh$. To this end we first show that F_t is in $\boldsymbol{B}$, the net $\{F_t\}$ converges to $P_T F$ under refinement with respect to the operator pseudonorm N_{HT}, and, for B in $\boldsymbol{B}$, $(I - B - F_t)^{-1}$ converges to $(I - B - F)^{-1}$ under refinement with respect to the pseudonorm N_{HT}. Convergence of solutions of $h = \Pi_t f + F_t h$ to the solution of $h = f + Fh$ under refinements follows as a corollary to this last statement. This result appears in the next section as Corollary 3.10 along with related computational considerations.

Theorem 3.6. The operator F_t is in $\boldsymbol{B}$. Furthermore, any variation function bk of F is also a variation function of F_t.

Proof. If bk is a variation function of F, $0 \leq t_{p-1} \leq u - c \leq v - c \leq t_p$, and f is in G then

$$
\begin{aligned}
|[F_t](v) - [F_t](u)| &= \{1/dk(t_{p-1}, t_p)\}\,|\,dk(t_{p-1}, v - c)\,[F\Pi_t f](t_p) + dk(v - c, t_p)[F\Pi_t f](t_{p-1}) \\
&\qquad - dk(t_{p-1}, u - c)\,[F\Pi_t f](t_p) - dk(u - c, t_p)[F\Pi_t f](t_{p-1})\,| \\
&\le b\{dk(u - c, v - c)/dk(t_{p-1}, t_p)\int_{[t(p-1),t(p)]} N_s(\Pi_t f)\, dk(s) \\
&\le b\, dk(u, v)\, N_{t(p)}(\Pi_t f) \\
&\le b\, dk(u, v)\, N_{t(p)}(f) \\
&\le b \int_{[u,v]} N_s(f)\, dk(s)
\end{aligned}
$$

If $0 \le t_{p-1} \le u - c \le t_p \le t_{q-1} \le v - c \le t_q$ then

$$
\begin{aligned}
|[F_t f](v) - [F_t f](u)| &\le |[F_t f](v) - [F_t f](t_{q-1} + c)| \\
&\qquad + \sum_{i=p+1}^{q-1} |[F_t f](t_i + c) - [F_t f](t_{i-1} + c)| + |[F_t f](t_p + c) - [F_t f](u)| \\
&\le b\, dk(u, t_p + c) N_{t(p)}(f) + b \sum_{i=p+1}^{q-1} dk(t_{i-1} + c, t_i + c) N_{t(i)}(f) \\
&\qquad + b\, dk(t_{q-1} + c, v) N_{t(q)}(f) \\
&\le b \int_u^v N_s(f) k\, d(s)
\end{aligned}
$$

and we are through. □

<u>Theorem</u> 3.7. The operator $P_T F$ is the limit through refinement of the net of finite dimensional operators $\{F_t\}$, t a partition of $[-r, T]$, with respect to the operator pseudonorm N_{HT}.

Let L and M denote the matrix representations of F and F_t, respectively. Here if $[u, v]$ is a subinterval of $[-r, T]$ then $M(v, u) = 0$ for $u \le c - r$ and $M(v, u) = [\Pi_t(P_T F)^* \Pi_T K(\,, u - c)](v)$ for $c - r \le u$. Equivalently $\langle M(v, u)e_j, e_i\rangle = \langle [\Pi_t P_T F \Pi_t K(\,, v)e_i](u - c), e_j\rangle$. One may show that Lemma 3.4 holds with M the matrix representation of F_t rather than for $\Pi_t F \Pi_t$.

Proof of Theorem 3.7. If $\{s_p\}_{0,m}$ is a partition of $[-r, T]$, $\{t_q\}_{0,n}$ is a refinement of s, $c = \text{mesh}(t)$, f is in G_H, x is in X, and u is in $[s_{p-1}, s_p]$ for some positive integer p then

$$
\begin{aligned}
|\langle [Ff](u) - [F_tf](u), x\rangle| &= |Q_H(f, L(\ , u)x - M(\ , u)x| \\
&\le N_H(P_Tf)[N_H(L(\ , u)x - L(\ , s_{p-1})x) + N_H(L(\ , s_{p-1})x \\
&\qquad - M(\ , s_{p-1} + c)x) + N_H(M(\ , s_{p-1} + c)x - M(\ , u)x)\} \\
&\le N_H(P_Tf)\{N_H(P_TF)\,|x|\,dk(s_{p-1}, u)^{1/2} \\
&\qquad + (N_H(L(\ , s_{p-1})x)^2 - \Sigma_t |dL(\ , s_{p-1})x|^2/dk)^{1/2} \\
&\qquad + N_H(P_TF)\,|x|\,(dk(s_{p-1}, u) + dk(0, c)^{1/2}\} \\
&\le N_H(P_Tf)\{3N_H(P_TF)\,|x|\,(\text{mesh}(s))^{1/2} \\
&\qquad + (N_H(L(\ , s_{p-1})x)^2 - \Sigma_t |dL(\ , s_{p-1})x|^2/dk)^{1/2}\}
\end{aligned}
$$

Suppose that $\varepsilon > 0$, $\{s_p\}_{p=0,n}$ is a partition of $[-r, T]$ such that $\text{mesh}(s) < (\varepsilon/6(N_H(P_TF) + 1))^2/d$, t' is a refinement of s such that if t refines t', $p = 1, 2, \cdots$, n, and $i = 1, 2, \cdots, d$ then $N_H(L(\ , s_{p-1})e_i)^2 - \Sigma_t |dL(\ , s_{p-1})e_i|^2/dk < (\varepsilon/2)^2/d$. If t refines t' and $|x| = 1$ then

$$
\begin{aligned}
|\langle [Ff](u) - [F_tf](u), x\rangle| &\le \sum_{i=1}^{d} |x_i|\,|\langle [Ff](u) - [F_tf](u), e_i\rangle| \\
&\le \sum_{i=1}^{d} |x_i|\,N_H(P_Tf)\varepsilon/d^{1/2} \\
&\le \varepsilon\,|x|\,N_H(P_Tf)
\end{aligned}
$$

i.e., $N_{HT}(F - F_t) \le \varepsilon$. □

Theorem 3.8. If B and F are in $\boldsymbol{B}$ then $(I - B - F)^{-1}$ is the limit through refinement of the net $\{(I - B - F_t)^{-1}\}$, t a partition of $[-r, T]$, with respect to the operator pseudo-norm N_{HT}.

Lemma 3.9. If bk is a variation function for F on $[0, T]$ then $be^{bT}k$ is a variation function for $(I - F)^{-1} - I$ on $[0, T]$.

Proof. If $h = f + Fh$ then by the Gronwall inequality $N_u(h) \leq e^{bu}N_u(f)$ for each u in $[0, T]$. Thus for each subinterval $[u, v]$ of $[0, T]$

$$|[(I - F)^{-1}f](v) - f(v) - [(I - F)^{-1}f](u) + f(u)|$$
$$\leq b\int_u^v N_s((I - F)^{-1}f)\, dk(s)$$
$$\leq b\int_u^v e^{bs}N_s(f)\, dk(s)$$
$$\leq be^{bT}\int_u^v N_s(f)\, dk(s) \quad \square$$

Proof of Theorem 3.8. Assume b_1k and b_2k are variation functions for B and F, respectively. Then $N_T((I - B - F_t)^{-1}) \leq 1 + be^{bT}\{k(T) - k(0)\}$, where $b = b_1 + b_2$. Hence, for each f in G_H,

$$N_T((I - B - F_t)^{-1}f - (I - B - F)^{-1}f)$$
$$\leq N_T((I - B - F_t)^{-1})N_{HT}(F - F_t)N_H(P_T(I - B - F)^{-1})N_H(P_Tf)$$
$$\leq (1 + be^{bT}\{k(T) - k(0)\})N_{HT}(F - F_t)N_H(P_T(I - B - F)^{-1})N_H(P_Tf)$$

i.e.,

$$N_{HT}((I - B - F_t)^{-1} - (I - B - F)^{-1})$$
$$\leq (1 + be^{bT}\{k(T) - k(0))\}N_H(P_T(I - B - F)^{-1})N_{HT}(F - F_t)$$

Thus $(I - B - D)^{-1}$ is the limit with respect to the operator pseudonorm N_{HT} of the net $\{(I - B - F_t)^{-1}\}$, t a partition of $[0, T]$. $\square$

4. Some Numerical Considerations

For a given system $h = f + Bh$ one expects B to be defined in terms of familar integrals and its matrix representation LB can be obtained using some existing numerical quadrature code. To a somewhat lesser extent one expects an input-output operator to be defined in terms of familar integrals. At this point we are not concerned with computing representations L(s, t) of system operators for various values of s and t in S but wish to illustrate the algebraic structure inherent in the approximations introduced in the previous section. It will be shown that these approximations can be written in a matrix form which facilitates the computation of operator approximations and approximate solutions to operator equations. The time invariance assumption allows further simplification of this structure and leads to a reduction in storage requirements necessary for the calculation of operator approximations.

If $\{t_p\}_{0,n}$ is an increasing sequence in S with $t_0 = -r$, $t_{p-1} \leq s \leq t_p$, f is in G, and F is in ***A*** or ***B*** then

$$[\Pi_t F \Pi_t f](s) = \{dk(s, t_p)[F\Pi_t f](t_{p-1}) + dk(t_{p-1}, s)[F\Pi_t](t_p)\}/dk(t_{p-1}, t_p)$$

Thus to evaluate $\Pi_t F \Pi_t$ one needs to know $[F\Pi_t f](t_p)$ for $p = 0, 1, \cdots, n$. As previously defined, f_t denotes the $(n + 1)d$ vector given by $f_t = (f(t_0)^*, f(t_1)^*, \cdots, f(t_n)^*)^*$. Let LF_t denote the $(n + 1)d \times (n + 1)d$ block matrix whose (p, q) block is given by $LF_t(p, q) = LF(t_p, t_q)$. As in Section 2, K_t is defined in a analogous way.

Recall $\Pi_t f = \Sigma_{q=0,n} K(\,, t_q)x_q$ where $x_q = [(K_t)^{-1} f_t](q)$ for $q = 0, 1, \cdots, n$. Hence for s in S $[F\Pi_t f](s) = \Sigma_{q=0,n}[FK(\,, t_q)](s)x_q = \Sigma_{q=0,n} L(t_q, s)^* x_q$ and, for $p = 0, 1, \cdots, n$,

$$\begin{aligned}
[F\Pi_t f](t_p) &= \sum_{q=0}^{n} [FK(\,, t_q)](t_p)x_q \\
&= \sum_{q=0}^{n} (LF(t_q, t_p)^*[(K_t)^{-1} f_t](q) \\
&= \sum_{q=0}^{n} (LF_t^*(p, q)[(K_t)^{-1} f_t](q) \\
&= [(LF_t)^*(K_t)^{-1} f_t](p) \qquad (3.11)
\end{aligned}$$

Note that $(LF_t)^*(p, q) = LF_t(q, p)^*$. Let us look more closely at the matrix $(LF_t)^*(K_t)^{-1}$.

Recall from Section 2 the form of the block symmetric tridiagonal matrix $(K_t)^{-1}$. First note that for $0 \le p \le n$

$$[(LF_t)^*(K_t)^{-1}](p, 0) = (LF_t)^*(p, 0)(K_t)^{-1}(0, 0) + (LF_t)^*(p, 1)(K_t)^{-1}(1, 0)$$

$$= LF_t(0, p)^*(K_t)^{-1}(0, 0) + LF_t(1, p)^*(K_t)^{-1}(1, 0)$$

$$= LF(t_0, t_p)^*(K_t)^{-1}(0, 0) + LF(t_1, t_p)^*(K_t)^{-1}(1, 0)$$

and for $0 < q < n$

$$[(LF_t)^*(K_t)^{-1}](p, q)$$

$$= (LF_t)^*(p, q - 1)(K_t)^{-1}(q - 1, q) + (LF_t)^*(p, q)(K_t)^{-1}(q, q)$$

$$+ (LF_t)^*(p, q + 1)(K_t)^{-1}(q + 1, q)$$

$$= LF_t(q - 1, p)^*(K_t)^{-1}(q - 1, q) + LF_t(q, p)^*(K_t)^{-1}(q, q)$$

$$+ LF_t(q + 1, p)^*(K_t)^{-1}(q + 1, q)$$

$$= LF(t_{q-1}, t_p)^*(K_t)^{-1}(q - 1, q) + LF(t_q, t_p)^*(K_t)^{-1}(q, q)$$

$$+ LF(t_{q+1,p})^*(K_t)^{-1}(q + 1, q)$$

For $0 \le p < q$ using (3.4) one obtains $[(LF_t)^*(K_t)^{-1}](p, q) = LF(t_p, t_p)^*\{(K_t)^{-1}(q - 1, q) + (K_t)^{-1}(q, q) + (K_t)^{-1}(q + 1, q)\} = 0$. Similarly $[(LF_t)^*(K_t)^{-1}](p, n) = 0$. Thus $(LF_t)^*(K_t)^{-1}$ is block lower triangular. If F is in $\boldsymbol{A}$ and $-r \le t_q \le t_p \le 0$ then $[(LF_t)^*(K_t)^{-1}](p, q) = I_d$ for $p = q$ and $[(LF_t)^*(K_t)^{-1}](p, q) = 0I_d$ for $p \ne q$. If F is in $\boldsymbol{B}$ and $-r \le t_q \le t_p \le 0$ then $[(LF_t)^*(K_t)^{-1}](p, q) = 0I_d$.

If F is time invariant and other simplifying assumptions are made then the structure of $(LF_t)^*(K_t)^{-1}$ can be further delineated. Let $k(t) = 1 + r + t$ for $-r \le t < 0$

and $k(t) = 1 + r + r_1 + t$ for $0 \leq t$. Also let $t_p = -r + pc$, $p = 0, 1, \cdots, n$ with $r = p'c$ for some positive integer p'. For $p = p' + 1$

$$[(Lf_t)^*(K_t)^{-1}](p' + 1, p' + 1)$$
$$= LF(t_{p'}, t_{p'+1})^*(-1/(r_1 + c)) + LF(t_{p'+1}, t_{p'+1})^*(1/(r_1 + c) + 1/c)$$
$$+ LF(t_{p'+2}, t_{p'+1})^*(-1/c)$$
$$= (1/(r_1+c)\{LF(t_{p'+1}, t_{p'+1})^* - LF(t_{p'}, t_{p'+1})^*\}$$

For $p > p' + 1$

$$[(LF_t)^*(K_t)^{-1}](p, p)$$
$$= LF(t_{p-1}, t_p)^*(-1/c) + LF(t_p, t_p)^*(2/c) + LF(t_p, t_p)^*(-1/c)$$
$$= (LF(t_p, t_p)^* - LF(t_{p-1}, t_p)^*)/c$$
$$= (LF(t_{p'+1}, t_{p'+1})^* - LF(t_{p'}, t_{p'+1})^*)/c$$
$$= \{LF_t(p' + 1, p' + 1)^* - LF_t(p', p' + 1)^*\}/c$$

Also for $p > p' + 1$

$$[(LF_t)^*(K_t)^{-1}](p, p' + 1)$$
$$= LF_t(p', p)^*(-1/(r_1 + c)) + LF_t(p' + 1, p)^*(1/(r_1) + 1/c)$$
$$+ LF_t(p' + 2, p)^*(-1/c)$$
$$= (1/(r_1 + c)\{LF_t(p' + 1, p)^* - LF_t(p', p)^*\}$$
$$- (1/c)\{LF_t(p' + 1, p)^* - LF_t(p' + 1, p)^*\}$$
$$= (1/(r_1 + c)\{LF_t(p' + 1, p)^* - LF_t(p', p)^*\}$$
$$- (1/c)\{LF_t(p' + 1, p - 1)^* - LF_t(p', p - 1)^*\}$$

and for $p' + 1 < q < p$ with $p = q + j$

$$[(LF_t)^*(K_t)^{-1}](p, q)$$
$$= LF(t_{q-1}, t_p)^*(-1/c) + LF(t_q, t_p)^*(2/c) + LF(t_{q+1}, t_p)^*(-1/c)$$
$$= (LF(t_q, t_p)^* - LF(t_{q-1}, t_p)^*)/c - (LF(t_{q+1}, t_p)^* - LF(t_q, t_p)^*)/c$$
$$= (LF(t_{p'+1}, t_{p'+1+j})^* - LF(t_{p'}, t_{p'+1+j})^*)/c$$
$$- (LF(t_{p'+1}, t_{p'+j})^* - LF(t_{p'}, t_{p'+j})^*)/c$$
$$= (LF_t(p' + 1, p' + 1 + j)^* - LF_t(p', p' + 1 + j)^*)/c$$
$$- (LF_t(p' + 1, p' + j)^* - LF_t(p', p' + j)^*)/c$$

For F in ***A*** it follows that $(L_t)^*(K_t)^{-1}$ may be written inpartitioned form as

$$\left[\begin{array}{c|c} I_{(p'+1)d\times(p'+1)d} & 0_{(p'+1)d\times(n-p')d} \\ \hline \boldsymbol{S}_{(n-p')d\times(p'+1)d} & \boldsymbol{T}_{(n-p')d\times(n-p')d} \end{array}\right]$$

where ***T*** is a block lower triangular.

For F in ***B*** $(L_t)^*(K_t)^{-1}$ may be written in partitioned form as

$$\left[\begin{array}{c|c} 0_{(p'+1)d\times(p'+1)d} & 0_{(p'+1)d\times(n-p')d} \\ \hline \boldsymbol{S}_{(n-p')d\times(p'+1)d} & \boldsymbol{T}_{(n-p')d\times(n-p')d} \end{array}\right]$$

where, again, ***T*** is a block lower triangular. In either case note that $\boldsymbol{T}(p, p)$ is constant for $p' + 2 \leq p$ and $T(p, p - j)$ is constant for $p' + 2 < p$ and $1 \leq j < p - 1$. If $r_1 = 0$ then ***T*** is block Toeplitz, i.e., constant blocks along diagonals.

As a corollary to Theorem 3.8 we now consider the convergence of solutions of $h = \Pi_t f + F_t h$ to the solution of $h = f + Fh$. Let M_t denote the $(n + 1)d \times (n + 1)d$ block matrix defined by

$$M_t(p, q) = \begin{cases} 0 & p = 0 \\ [(LF_t)^*(K_t)^{-1}](p - 1, q) & 1 \leq p \leq n,\ 0 \leq q \leq n \end{cases}$$

Here the first row of the partitioned matrix M_t is a row of zeros, while the p-th row of M_t is the (p - 1)-th row of $(L_t)^*(K_t)^{-1}$ for $p = 2, \cdots, n$.

If $h = \Pi_t f + F_t h$ then h is in G_H and, for $0 \leq p \leq n$, $h(t_p) = f(t_p) + [F_t](t_p)$ or $h_t(p) = f_t(p) + [M_t h_t](p)$, i.e., $h_t = f_t + M_t h_t$ or $h_t = (I_{(n+1)d} - M_t)^{-1} f_t$.

<u>Corollary</u> 3.10. If f is in G_H, $h = (I - F)^{-1}f$, and ε is a positive number then there is a partition s of [0, T] such that if $\{t_p\}_{0,n}$ refines s and $0 \leq p \leq n$ then $| h(t_p) - [(I_{(n+1)d} - M_t)^{-1} f_t](p) | < \varepsilon$.

<u>Proof</u>. Using Theorem 3.8 with B = 0, There is a partition s of [0, T] such that if $\{t_p\}_{0,n}$ refines s then $N_{HT}((I - F)^{-1} - (I - F_t)^{-1})$ is smaller than 1 and $(1/2)\varepsilon/(N_H(f) + 1)$ and $N_H(f - \Pi_t f) < (1/2)\varepsilon/(N_{HT}(I - F)^{-1}) + 1)$. Thus

$$
\begin{aligned}
&N_T((I - F)^{-1} f - (I - F_t)^{-1} \Pi_t f) \\
&\quad \leq N_{HT}((I - F)^{-1} - (I - F_t)^{-1}) N_H(f) + N_{HT}((I - F_t)^{-1}) N_H(f - \Pi_t f) \\
&\quad \leq (1/2)\varepsilon + (N_{HT}((I - F)^{-1}) + 1) N_H(f - \Pi_t f) \\
&\quad \leq \varepsilon
\end{aligned}
$$

Since $[(I - F_t)^{-1} \Pi_t f](t_p) = (I_{(n+1)d} - M_t) f_t$ we are through. □

As a consequence of this corollary, one can obtain an approximate solution to the equation h = f + Fh by forming the K-polygon which agrees with the solution of the matrix equation $h_t = f_t + M_t h_t$ at the mesh points.

<u>Some numerical examples</u>. If h = Ff then equation (3.11) may be written as $h_t = (LF_t)^*(K_t)^{-1} f_t$. For some basic operators which frequently appear in applications the structure of the matrix $(LF_t)^*(K_t)^{-1}$ is particularly simple. Let $k(t) = 1 + r + t$ and $t_p = \{-r + pc\}_{p=0,n}$ be a partition of [-r, T] with $r = p'c$ for some positive integer p′.

If

$$
[Ff](t) = \begin{cases} 0 & -r \leq t \leq 0 \\ \alpha \int_0^t f(\tau) d\tau & 0 \leq t \end{cases}
$$

then $(LF_t)^*(K_t)^{-1}$ =

$$
\begin{array}{cc}
 & \begin{array}{ccccccccc} 0 & p' & p'+1 & & & & & & \end{array} \\
\begin{array}{c} p' \\ p'+1 \\ \\ \\ \\ \\ \\ \end{array} &
\left[\begin{array}{cc|ccccccc}
0 & 0 & 0 & 0 & \cdot & \cdot & \cdot & & \\
\hline
0 & \alpha\frac{c}{2} & \alpha\frac{c}{2} & & & & & & \\
0 & \alpha\frac{c}{2} & \alpha c & \alpha\frac{c}{2} & & & & & \\
0 & \cdot & \cdot & \cdot & \cdot & & & & \\
0 & \cdot & \cdot & \cdot & \cdot & \cdot & & & \\
0 & \cdot & \cdot & \cdot & \cdot & \cdot & \cdot & & \\
0 & \alpha\frac{c}{2} & \alpha c & \alpha c & \cdot & \cdot & \cdot & \alpha c & \alpha\frac{c}{2}
\end{array}\right]
\end{array}
$$

If

$$
[Ff](t) = \begin{cases} 0 & -r \le t \le 0 \\ \alpha\int_0^t h(\tau - r)d\tau & 0 \le t \end{cases}
$$

then $(LF_t)^*(K_t)^{-1}$ =

$$
\begin{array}{cc}
 & \begin{array}{cccccccccc} 0 & & & & & & & & & \end{array} \\
\begin{array}{c} 0 \\ p' \\ p'+1 \\ \\ \\ \\ \\ \\ \\ \\ \end{array} &
\left[\begin{array}{cccccccccc}
0 & 0 & 0 & \cdot & \cdot & \cdot & & & & \\
\hline
\alpha\frac{c}{2} & \alpha\frac{c}{2} & & & & & & & & \\
\alpha\frac{c}{2} & \alpha c & \alpha\frac{c}{2} & & & & & & & \\
\alpha\frac{c}{2} & \alpha c & \alpha c & \alpha\frac{c}{2} & & & & & & \\
\alpha\frac{c}{2} & \alpha c & \alpha c & \alpha c & \alpha\frac{c}{2} & & & & & \\
\cdot & \cdot & \cdot & \cdot & \cdot & \cdot & & & & \\
\cdot & \cdot & \cdot & \cdot & \cdot & & \cdot & & & \\
\cdot & \cdot & \cdot & \cdot & \cdot & & & \cdot & & \\
\alpha\frac{c}{2} & \alpha c & \alpha c & \alpha c & \cdot & \cdot & \cdot & \cdot & \alpha c & \alpha\frac{c}{2}
\end{array}\right]
\end{array}
$$

If

$$[Ff](t) = \begin{cases} 0 & -r \le t \le 0 \\ \alpha\int_0^t (t-\tau)f(\tau)d\tau & 0 \le t \end{cases}$$

then $(LF_t)^*(K_t)^{-1} =$

$$\begin{array}{c} \\ 0 \\ p' \\ \\ p'+1 \\ \\ \\ \\ \\ \\ \\ p'+j \end{array}
\left[\begin{array}{cc:cccccc}
0 & p' & p'+1 & & & & & \\
0 & 0 & 0 & 0 & \cdot & \cdot & \cdot & \\
\hdashline
0 & \alpha(\frac{1}{2}-\frac{1}{6})c^2 & \alpha\frac{c^2}{6} & & & & & \\
0 & \alpha(1-\frac{1}{6})c^2 & \alpha c^2 & \alpha\frac{c^2}{6} & & & & \\
0 & \alpha(\frac{3}{2}-\frac{1}{6})c^2 & 2\alpha c^2 & \alpha c^2 & \alpha\frac{c^2}{6} & & & \\
 & & & & & & & \\
0 & \alpha(\frac{j}{2}-\frac{1}{6})c^2 & (j-1)\alpha c^2 & (j-2)\alpha c^2 & \cdots & 2\alpha c^2 & \alpha c^2 & \alpha\frac{c^2}{6}
\end{array}\right]$$

These matrices have been determined using the identities derived previously and exact calculation of $[FK(\ ,t_p)](t_q)$ for $p = 0, 1, \cdots, p'+1$ and $q = 0, 1, \cdots, n$. Note in the first and second examples the approximation scheme has resulted in the use of the trapezoid rule to approximate the integrals. Complexities which arise if the delay parameter in the second example is not a multiple of the step size will be discussed in the section on parameter estimation.

In order to convey some feeling for the convergence of operator approximations we have applied the discretizations to some particular examples.

<u>Example</u> 3.2. Consider the system $h = f + Bh$ with B as in Example 3.1, $r = 1$, $\alpha = 1$, $k(t) = 2 + t$, and $f(t) = \sin 2\pi t$ for $-1 \le t$. The exact solution, computed by stepwise integration, is

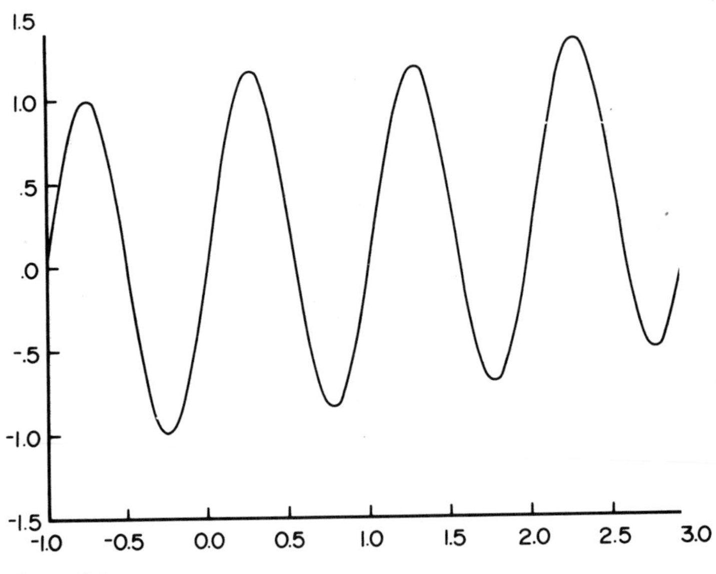

Figure 3.1

$$
h(t) = \begin{array}{ll}
\sin 2\pi t - (1/2\pi)\{\cos 2\pi(t-1)\} & 0 \leq t \leq 1 \\
\sin 2\pi t - (1/2\pi)\{\cos 2\pi(t-1) - 1\} - (1/2\pi)^2 \sin 2\pi(t-2) + 1/2\pi(t-1) & 1 \leq t \leq 2 \\
\sin 2\pi t - (1/2\pi)\{\cos 2\pi(t-1) - 1\} - (1/2\pi^2 \sin 2\pi(t-2) + 1/2\pi(t-1) & 2 \leq t \leq 3 \\
\quad + (1/2\pi)^3 \{\cos 2\pi(t-3) - 1\} + \{(1/2\pi)(t-2)\}^{-2} &
\end{array}
$$

and is depicted graphically in Figure 3.1. Let $A = (I - B)^{-1}$ and $t_p = -1 + pc$ for $p = 0, 1, \cdots, n$ and $1 = p'c$ for some positive integer p'. Using exact calculations of $[AK(, t_p)](t)$ for $p = 0, 1, \cdots, p' + 1$ and the time invariance property, see pages 98-100, we have calculated $h_t = (LA_t)^*(K_t)^{-1}f_t$ for various choices of the step size c.

The error of the approximation is indicated in Table 3.1.

The error resulting from the approximation $h_t = (I - M_t)^{-1}f_t$ of Corollary 3.10 is indicated in Table 3.2.

Note the quadratic convergence of the first approximation as compared to the linear convergence of the second approximation.

Table 3.1

c	$\max \mid h(t_i) - [\Pi_t A \Pi_t f](t_i) \mid$	
1/8	.29704	10^{-1}
1/16	.7498	10^{-2}
1/32	.1871	10^{-2}
1/64	.46752	10^{-3}
1/128	.11687	10^{-3}
1/256	.29218	10^{-4}

Table 3.2

c	$\max \mid h(t_i) - h_t(i) \mid$	
1/8	.17239	
1/16	.87638	10^{-1}
1/32	.44616	10^{-1}
1/64	.22449	10^{-1}
1/128	.11252	10^{-1}
1/256	.56317	10^{-2}

Example 3.3. Consider the Volterra integral equation

$$h(t) = f(t) + \int_0^t e^{\alpha(t-s)} h(s)\, ds$$

with $f(t) = \cos 2\pi t$ for $0 \leq t$ and $\alpha = -3$. Here $[Bh](t) = \int_{[0,t]} e^{\alpha(t-s)} h(s)\, ds$ and $r = 0$. The exact solution, obtained by solving an equivalent system of ordinary differential equations [12, 23], is

$$h(t) = \cos 2\pi t + (1/2)(1 + \pi^2)^{-1} \{\pi \sin 2\pi t + 2\cos 2\pi t - e^{-2t}\}$$

see Figure 3.2.

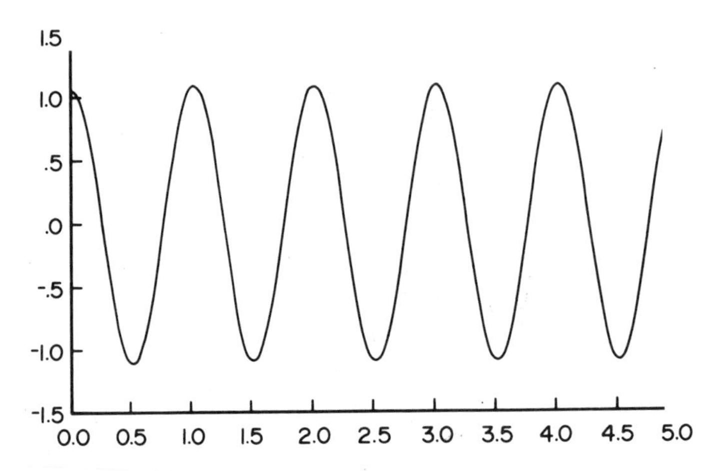

Figure 3.2

Let $A = (I - B)^{-1}$ and $t_p = pc$, $p = 0, 1, \cdots, n$, be a partition of the interval [0, 5]. Exact calculation of $[AK(, 0)](t)$ and $[AK(, 5)](t)$, $0 \le t \le 5$ and use of the time invariance property allow the calculation of the approximation $h_t = (LA_t)^*(K_t)^{-1}f_t$ for various choices of the stepsize c. The error of the approximation is indicated in Table 3.3.

5. System Identification

The term <u>system identification problem</u> is used to refer to those problems which require finding or estimating the parameters of a given representation of the

Table 3.3

c	$\max \mid h(t_i) - [\Pi_t A \Pi_t f](t_i) \mid$	
1/8	.80389	10^{-2}
1/16	.20942	10^{-2}
1/32	.52304	10^{-3}
1/64	.13087	10^{-3}
1/128	.32733	10^{-4}
1/256	.81832	10^{-6}

system using input-output information. Frequently, the representation must be refined to a point where an algorithm can be applied. The task of refinement using system properties such as linearity or time invariance, either as an assumption or as a description of the system, will be thought of as part of the system identification problem. In this section, we consider the problem of "identifying" an operator F in ***A*** or ***B*** given pairs (f, h) in F. For the functional equation

$$h(t) = \begin{cases} f(t) & -r \le t \le 0 \\ f(t) + \int_0^t \alpha h(\tau - \rho)d\tau & 0 \le t \le T \end{cases}$$

a representation of the input-output operator A was given in Example 3.1. The Riesz Representation Theorem tells us there is a function g from $S \times S$ to the reals such that, for each t in S, g(t,) has bounded variation and $[Af](t) = \int_{[-r,t]} f(s)dg(t, s)$ for each f in G. The system identification problem in this context consists of finding g using input-output information. Since we are interested in hereditary systems described by operators F in ***A*** or ***B***, and the identification problem for such systems can be posed as the problem of finding the matrix representation of the operator given pairs (f, h) in F.

In general, the solution of an identification problem requires extensive knowledge of input-output pairs (f, h). Remember if F is in ***A*** or ***B*** then $LF(u, v)_{i,j} = \langle [FK(, u)e_i](v), e_j \rangle$, i.e., the representation of F is determined from knowledge of $FK(, u)e_i$ for $i = 1, \cdots, d$ and u in S. Note that $[Ff](t) = [FP_0f](t) + [F(I - P_0)f](t)$ and so the identification problem can be decomposed into the separate problems of identifying F on P_0G_H and on $(I - P_0)G_H$, respectively.

To identify F restricted to P_0G_H requires extensive input-output information or some approximations. Here $[Ff](t) = LF(-r, t)*f(-r) + \int_{[-r,0]} dLF(s, t)*df(s)/dk$ for f in P_0G_H and $0 \le t$ and we need to know $LF(u, t)_{i,j} = \langle [FK(, u)e_i](t), e_j \rangle$ for $-r \le u \le 0$ and $0 \le t$. Recall from (3.4) that LF(u, t) is known for $-r \le u \le 0$ and $-r \le t \le 0$. However, without complete knowledge of [FK(, u)](t) for $-r \le u \le 0$ and $t > 0$ we are unable to compute the matrix representation of F restricted to P_0G_H. On the other hand, if $s = \{s_p\}_{p=0,n}$ is a partition of [-r, 0] and $[FK(, s_p)](t)$ is known for $-r \le t$ and s_p in s then one can compute $[\Pi_s F \Pi_s f](t)$ for $-r \le t$ whenever f is in P_0G_H. Thus the approximation $\Pi_s F \Pi_s$ to F on P_0G_H can be computed using limited data.

In this section we are primarily concerned with the problem of identifying the input-output operator under the assumption that the system is initially at rest. That is we wish to obtain the representation of F restricted to $(I - P_0)G_H$. If f is in $(I - P_0) \cdot G_H$, then $[Ff](t) = \int_{[0,t]} dLF(s, t)*df(s)/dk(s)$ for $0 \leq t$ and consequently one need only compute LF(u, v) for $0 \leq u, v$. For the important time invariant operators this identification problem can be solved using limited data.

As a consequence of the time invariant assumption, by setting $b = s$, $u = 0$, $v = w = t - s$ in equation (3.5) one obtains the identity

$$LF(s, t) = LF(t, t) - LF(t - s, t - s) + LF(0, t - s) \tag{3.12}$$

for $0 \leq s \leq t$. Also, from the defining properties of LF (3.2)

$$LF(0, t)_{ij} = <(FK[\ , 0]e_i)(t), e_j>$$

and

$$LF(t, t)_{ij} = LF(T, t)_{ij} = <[FK(\ , T)e_i](t), e_j>$$

for $0 \leq t \leq T$. Thus, letting $f_{1i}(t) = K(t, 0)e_i$, $f_{2i}(t) = K(t, T)e_i$, $h_{1i}(t) = [Ff_{1i}](t)$ and $h_{2i}(t) = [Ff_{2i}](t)$ for $i = 1, \cdots, d$ and $-r \leq t$ the following results are readily obtained.

<u>Theorem</u> 3.11. The representation LF of a time invariant operator F in $\boldsymbol{A}$ or $\boldsymbol{B}$ is determined for $0 \leq t \leq T$ from knowledge of the 2d pairs $(f_{1i}(t), h_{1i}(t))$, $(f_{2i}(t), h_{2i}(t))$, $i = 1, \cdots, d$ and $0 \leq t \leq T$.

For $c > 0$, let $s = \{s_p\}_{p=0,n}$ be a patition or $[-r, T]$ with $s_p = -r + pc$. The matrix representation of $\Pi_s F \Pi_s$ agrees with LF at partition points (s_p, s_q) and so we obtain the following corollary.

<u>Corollary</u> 3.12. The matrix representation of the approximate operator $\Pi_s F \Pi_s$ is

determined for $0 \le s_p$, $s_q \le T$ by the 2d discrete input-output pairs $(\Pi_s f_{1i}(s_p)$, $\Pi_s h_{1i}(s_p))$, $(\Pi_s f_{2i}(s_p)$, $\Pi_s h_{2i}(s_p))$, $i = 1, \cdots, d$ and $p = 0, 1, \cdots, n$.

Knowledge of arbitrary pairs (f, h) in F is not sufficient to determine the representation of LF. However, Equation (3.12) may be used to determine the representation from more general information than indicated in Theorem 3.11.

In the time invariant case, if $\{s_p\}_{p=0,n}$ is given by $s_p = -r + pc$ for $p = 0, \cdots, n$ with $c > 0$ and $r = p'c$ for some integer p' and $k(t) = 1 + r + t$ for $r \le t$ then for f in $(I - P_0)G_H$ equation (3.10) becomes

$$<\Pi_s F \Pi_s f](c), x> = <f(c), \{LF(c, c) - (1/c)(LF(c, c) - LF(0, c))\}x>$$

and for $p > 1$

$$<\Pi_s F \Pi_s f](pc), x> = \sum_{q=1}^{p-1} <f(qc), \{LF(c, (p - q + 1)c)x - LF(0, (p - q + 1)c)x\}/c$$

$$- \{(LF(c, (p - q)c)x - LF(0, p - q)c)x\}/c>$$

$$+ <f(pc), \{LF(c, c)x - LF(0, c)x\}/c>$$

for each f in G_H, x in X. Thus for numerical problems only the first two rows of the discretization of LF, i.e., L(0,) and L(c,), are needed. Furthermore, equation (3.12) yields $(LF(c, pc) = LF(pc, pc) + LF(0, (p - 1)c) - LF((p - 1)c, (p - 1)c)$, for $p > 0$, i.e., the "row" LF(c,) may be calculated from the "diagonal" on LF.

In order to illustrate these ideas for some simple examples we introduce a simplification of (3.2) which is available for these examples. For each t in [-r, T] let $LF_1(, t)$ denote, if it exists, the derivative of the restriction of LF(, t) to [-r, t].

Theorem 3.14. If for each t in $(-r, \infty)$ the restriction of LF(, t) to [0, t] is differentiable and $LF_1(0,)$ is either continuous or has bounded variation then $<[Ff](t), x> = \int_{[0,t]} <df, LF_1(0, t - I)x>$ for each f in $(I - P_0)G_H$, x in X, and t in S.

Proof. If $0 \le s \le t$ and $0 \le c \le t - s$ then using the time invariance property,

$$LF(s + c, t) - L(s, t) = LF(t, t) - LF(t - s - c, t - s - h) + LF(0, t - s - c)$$
$$- \{LF(t, t) - LF(t - s, t - s) + LF(0, t - s)\}$$
$$= -LF(t - s - c, t - s - c) + LF(0, t - s - c)$$
$$+ LF(t - s, t - s) - LF(0, t - s)$$
$$= - \{LF(t - s, t - s) - LF(c, t - s)\} + \{LF(t - s, t - s) - LF(0, t - s)\}$$
$$= LF(c, t - s) - LF(0, t - s)$$

and so $LF_1(s, t) = LF_1(0, t - s)$. Suppose $\{s_p\}_{0,m}$ is a partition of [0, t] and f is in $(I - P_0)G_H$. There is a $d \times m$ matrix τ such that $s_{p-1} < \tau_{ip} < s_p$ for $i = 1, 2, \cdots, d$, $p = 1, 2, \cdots, m$ and for $j = 1, \cdots, d$

$$\sum_{p=1}^{m} <df(s_{p-1}, s_p), (LF(s_p, t) - LF(s_{p-1}, t))e_j>/dk(s_{p-1}, s_p)$$
$$= \sum_{p=1}^{m} \sum_{i=1}^{d} (df(s_{p-1}, s_p))_i (LF_1(\tau_{ip}, t))_{ij}$$
$$= \sum_{p=1}^{m} \sum_{i=1}^{d} (df(s_{p-1}, s_p))_i (LF_1(0, t - \tau_{ip}))_{ij}$$

Since $<[Ff](t), e_j> = \int_{[0,t]} <df, dLF(, t)e_j>/dk$ we obtain $<[Ff](t), e_j> = \int_{[0,t]} <df, LF_1(0, t - I)e_j>$ and the theorem follows. □

In the following exercise one is asked to show that standard integral representations of some simple operators in ***B*** can be identified from limited input-output data.

Exercise. Refer to parts a), b), and c), respectively, in the exercise in Section 1.

a) Show that $LB_1(0, t) = \alpha^* t$ and for f in $(I - P_0)G_H$ $<[Bf](t), x> = \int_{[0,t]} <df, LB_1(, t - I)x> = <\alpha\int_{[0,t]} fdI, x>$ and thus recover [Bf](t) exactly.
b) Show that $LB_1(0, t) = \alpha^* t^2/2$ and [Bf](t) can be recovered exactly for f in $(I - P_0)G_H$.

c) Show that $LB(0, t) = 0$ for $0 \le t \le \rho$, $LB_1(0, t) = \alpha^*(t - \rho)$ for $\rho \le t$, and $[Bf](t)$ can be recovered exactly for f in $(I - P_0)G_H$.

Example 3.4. To further illustrate these ideas we expand on Example 3.1. Assuming LA(0, t) and LA(t, t) can be measured exactly, one obtains

$$
LA(0,t)^* = \begin{cases}
(1 + r + t)I_d & -r \le t \le 0 \\
(1 + r)I_d + (\alpha/2)\{(1 + r + t - \rho)^2 - (1 + r - \rho)^2\}(1 + r)I_d & 0 \le t \le \rho \\
\quad + (\alpha/2)\{(1 + r)^2 - (1 + r - \rho)^2 + 2(1 + r)(t - \rho)\} + (\alpha^2)(1/2)\{t(1+r+t-2\rho)^2 & \\
\quad - t(1+r-\rho)^2 - t^2(1+r-2\rho) + \rho^2(1+r-2\rho) - (2/3)t^3 + (2/3)\rho^3\} & \rho \le t \le 2\rho
\end{cases}
$$

and

$$
LA(t,t)^* = \begin{cases}
(1 + r + t)I_d & -r \le t \le 0 \\
(1 + r + t)I_d + (\alpha/2)\{(1 + r + t - \rho)^2 - (1 + r - \rho)^2\} & 0 \le t \le \rho \\
(1 + r + t)I_d + (\alpha/2)\{(1 + r + t - \rho)^2 - (1 + r - \rho)^2\} & \\
\quad + (\alpha^2 t/2)\{(1 + r + t - 2\rho)^2 - (1 + r - \rho)^2\} & \\
\quad - \alpha^2\{(1 + r - 2\rho)(t^2/2 - \rho^2/2) + t^3/3 - \rho^3/3\} & \rho \le t \le 2\rho
\end{cases}
$$

and LA(s, t) can be determined for $0 \le s \le t \le 2\rho$ using (3.13). It follows that

$$
LA_1(s,t) = \begin{cases}
0I_d & t \le s \\
I_d & t - \rho \le s \le t \\
I_d + \alpha^*(t - s - \rho) & s \le t - \rho
\end{cases}
$$

Therefore for f in $(I - P_0)G_H$ one obtains for $0 \le s \le t \le \rho$

$$<[Af](t), e_j> = \int_0^t <df, L_1(0, t - I)e_j> = <f(t) - f(0), e_j> = <f(t), e_j>$$

and for $\rho \leq s \leq 2\rho$

$$<[Af](t), e_j> = \int_0^{t-\rho} <df, L_1(0), t - I)e_j> + \int_{t-\rho}^{t} <df, L_1(0, t - I)e_j>$$

$$= \int_0^{t-\rho} <df, (e_j + \alpha^*(t - I - \rho)e_j> + \int_{t-\rho}^{t} <df, e_j>$$

$$= <f(t - \rho), e_j> + \int_0^{t-\rho} <df, \alpha^*(t - I - \rho)e_j> + <f(t) - f(t - \rho), e_j>$$

$$= <f(t) + \alpha\int_0^{t} f(\tau - \rho)\, d\tau,\ e_j>$$

Thus for f in $(I - P_0)G_H$ we obtain $[Af](t) = f(t)$ for $0 \leq t \leq \rho$ and $[Af](t) = f(t) + \alpha\int_{[0,t]} f(\tau - \rho)\, d\tau$ for $\rho \leq t \leq 2\rho$. Theorem 3.12 implies that the matrix representation of the input-output operator is determined by 2d input-output pairs. In this example, we have illustrated that the representation of the input-output operator given in Example 3.1 can be determined exactly from knowledge of 2d input-output pairs.

Example 3.5. Consider the system

$$h(t) = f(t) - \int_0^t (t - s)\{h(s) + h(s - 1)\}\, ds - \int_0^t \{h(s) + h(s - 1/16)\}ds$$

for $t \geq 0$ assuming $h(t) = f(t)$ for $-1 \leq t \leq 0$. Differentiating twice leads to an equivalent second order delay equation

$$h''(t) = g(t) - h(t) - h(t - 1) - h'(t) - h'(t - 1/16)$$

Table 3.4

$f(t)\quad t \geq 0$	$\max \mid h(t_i) - [\Pi_t A \Pi_t f](t_i) \mid$	
t^2	.31161	10^{-3}
t^3	.24550	10^{-2}
$\sin 2\pi t$	.19373	10^{-2}

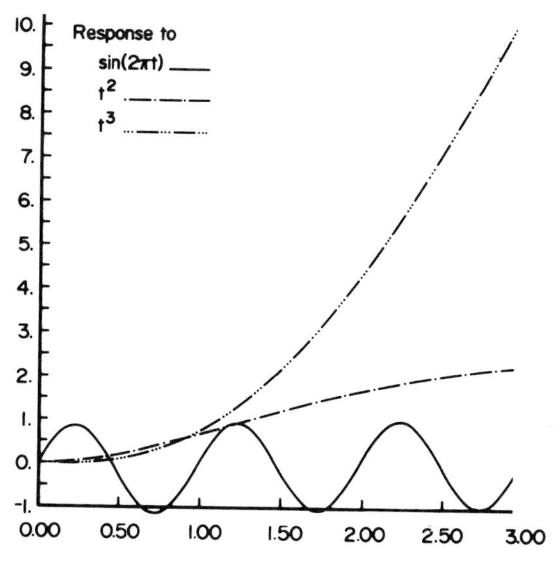

Figure 3.3

for $t \geq 0$ assuming $f(t) = h(0) + t(h'(0) + h(0) + h(-1/16)\} + \int_{[0,t]} (t - s)g(s)\, ds$ for $t \geq 0$. As a numerical experiment, the software package [25] was used to obtain the response of the system to the inputs $f(t) = K(t, 0)$ and $f(t) = K(t, 3)$ with $k(t) = 2 + t$ at discrete times $t_p = -1 + p/32$, $p = 0, 1, \ldots, 128$. By Corollary 3.13 the approximate input-output operator can then be determined. In Table 3.4 we indicate the difference between the responses obtained from the approximate input-output operator $\Pi_t A \Pi_t f$ and from the software package [25] to various inputs f in $(I - P_0)G_H$. The system response to these inputs is depicted in Figure 3.3.

6. Parameter Estimation

For the system $h = f + B(\gamma)h$ with $B(\gamma)$ a parameterized operator the parameter estimation problem could be phrased as follows: given an input-output pair (f', h') satisfying $h' = f' + B(\gamma')h'$ with γ' unknown, determine a scheme based upon the given observations and knowledge of the system operator $B(\gamma)$ to estimate the parameter γ'. If one could solve the system $h = f' + B(\gamma)h$ exactly in terms of the parameters to obtain the solution $h(\gamma)$ then γ' could be determined by adjusting γ so that $h(\gamma) = h'$ or equivalently $N_H(h(\gamma) - h') = 0$. We assume $h(\gamma_1) \neq h(\gamma_2)$ whenever $\gamma_1 \neq \gamma_2$.

Due to limitations in data measurements and in obtaining explicit solutions of the equation $h = f' + B(\gamma)h$ this scheme probably will not be supported by the data. That is measurements of the input f' and response h' are frequently only available at a finite number of sample points and one is only able to obtain an approximation $h_a(\gamma,)$ to the solution of the equation $h = f' + B(\gamma)h$.

With these considerations in mind it is necessary to formulate an algorithm based upon a reasonable amount of information which yields an estimate of the unknown parameter γ'. Assume that $B(\gamma)$ is known for each γ lying in an open subset Γ of R^m and we can calculate an approximate solution $h_a(\gamma,)$ of the equation $h = f + B(\gamma)h$ for each f. Also assume that the input f' and the response h' of $h = f' + B(\gamma')h$ are sampled at times $s = \{s_p\}_{p=0,n}$ with $-r = s_0 < s_1 < \cdots < s_n = T$ to obtain observations $\{f'(s_p), h'(s_p)\}_{p=0,n}$. An estimate for γ' can be obtained by minimizing an error functional $J(\gamma) = \| e(\gamma) \|$ where $e(\gamma)$ is an error vector and $\| \cdot \|$ denotes an appropriate norm. The error vector is assumed to have components $e(\gamma)_p = h_a(\gamma, s_p) - h'(s_p)$ for $p = 0, 1, \cdots, n$.

Solution of the parameter estimation problem requires that an optimization scheme be used to minimize the error functional. The central issue in the parameter estimation problem is how to choose a suitable approximation $h_a(\gamma,)$ to the true solution and an appropriate error functional $J(\gamma)$. In addition, analysis of the convergence of approximate solutions and their dependence upon system parameters is required.

In this section the operator approximations previously introduced and the given observations are used to define two approximate solutions to be used in the algorithm for estimating γ'. It is shown that approximate solutions $h_a(\gamma,)$ converge to the true solution h' as the sample points become dense in the sampling interval $[-r, T]$ and as the parameter γ converges to the true parameter γ'. It is also shown that use of an appropriate error functional yields a vector formula suitable for numerical optimization procedures. For low order delay differential equations, dependence of these approximate solutions upon system parameters becomes explicit and many computational simplifications arise.

In Section 1 a variation k was specified and on the interval $[0, T]$ and a positive number b was associated with each operator in $\boldsymbol{B}$. Here it is assumed that the inequalities in the definition of $\boldsymbol{B}$ hold for $B(\gamma)$ with a fixed constant b for all γ in Γ. It is also assumed that $B(\gamma)$ is continuous in γ with respect to the operator norm N_T.

The observations $\{(f'(s_p), h'(s_p))\}_{p=0,n}$ are assumed to arise from a fixed input-output pair (f', h') in $G_H \times G_H$ corresponding to a parameter γ'. Recall that $(f')_s$ and $(h')_s$ denote the vectors in $R^{d(n+1)}$ such that $(f'_s)(p) = f'(s_p)$ and $(h'_s)(p) = h'(s_p)$. Also $\theta_s(f')_s$ and $\theta_s(h')_s$ denote the K-polygonal functions which agree with the observations $f'(s_p)$ and $h'(s_p)$ for $p = 0, 1, \cdots, n$. Thus our observations are represented by a pair $\{(f')_s, (h')_s\}$. A first approximation to h' is defined by

$$h_1(\gamma, s) = \theta_s(f')_s + \Pi_s B(\gamma)\theta_s(h')_s.$$

Clearly $h_1(\gamma, s)$ is in $\Pi_s G_H$. Further h_1 approximates h' with respect to the norm N_T.

Theorem 3.15. For $\varepsilon > 0$ there exists a partition t of $[-r, T]$ and a $\delta > 0$ such that if s refines t and γ is in Γ with $|\gamma - \gamma'| < \delta$ then $N_T(h_1(\gamma, s) - h') < \varepsilon$.

Proof. Recall that $\theta_s(h')_s = \Pi_s h'$. If γ in Γ and s a partition of $[-r, T]$ then

$$\begin{aligned}
&N_T(h_1(\gamma, s) - h') \\
&\quad \le N_T(\theta_s(f')_s - f') + N_T(\Pi_s B(\gamma)\Pi_s h' - \Pi_s B(\gamma')\Pi_s h') \\
&\qquad + N_T(\Pi_s B(\gamma')\Pi_s h' - B(\gamma')h') \\
&\quad \le N_T(\theta_s(f')_s - f') + N_T(B(\gamma) - B(\gamma'))N_T(\Pi_s h') \\
&\qquad + N_{HT}(\Pi_s B(\gamma')\Pi_s - P_T B(\gamma'))N_H(h')
\end{aligned}$$

For $\varepsilon > 0$ choose a partition t of $[-r, T]$ such that if s is a refinement of t then $N_T(\theta_s(f')_s - f') < \varepsilon/3$, and, using Theorem 3.2, $N_{HT}(\Pi_s B(\gamma')\Pi_s - P_T B(\gamma')) < \varepsilon/(3(N_H(h') + 1))$. Also choose $\delta > 0$ so that $N_T(B(\gamma) - B(\gamma')) \le \varepsilon/(3(N_T(h') + 1))$ whenever $|\gamma - \gamma'| < \delta$. Since $N_T(\Pi_s h') \le N_T(h') + 1$ the result follows. □

A second approximate solution is motivated by the following considerations. Let $A(\gamma) = (I - B(\gamma))^{-1}$ for each γ in Γ. Then $h' = A(\gamma')f'$ and we might choose $h_2(\gamma, s) = \Pi_s A(\gamma)\Pi_s f'$ as a second approximate solution. However, a simple parameterized representation of $\Pi_s A(\gamma)\Pi_s$ is difficult to obtain. Since $A(\gamma) = (I - B(\gamma))^{-1}$ an alternate approach is to approximate $I - B(\gamma)$. Let $B_s(\gamma)$ be as defined in equation (3.10), the second approximate solution to h´ is defined by

$$h_2(\gamma, s) = (I - B_s(\gamma))^{-1}\theta_s(f')_s$$

or equivalently as the solution of the equation

$$h_2(\gamma, s) = \theta_s(f')_s + B_s(\gamma)h_2(\gamma, s).$$

The inverse of $I - B_s(\gamma)$ exists by Theorem 3.6.

Theorem 3.16. For each $\varepsilon > 0$ there exists a partition t of [-r, T] and a $\delta > 0$ such that if s refines t and γ in Γ satisfies $|\gamma - \gamma'| < \delta$ then $N_T(P_Th_2(\gamma, s) - P_Th') < \varepsilon$.

Proof. By Theorem 3.6, bk is a variation for $B_s(\gamma)$ for all γ in Γ and hence $N_T((I - B_s(\gamma))^{-1}f) \leq N_T(f)\, e^{b(k(T)-k(0))}$ for all f in G_H, see Chapter 1. Using the triangle inequality one obtains

$$\begin{aligned} N_T(h_2(\gamma, s) - h') &\leq N_T((I - B_s(\gamma))^{-1}(\theta_s(f')_s - f')) + N_T((I - B_s(\gamma))^{-1}f' \\ &\quad - (I - B_s(\gamma'))^{-1}f') + N_T((I - B_s(\gamma'))^{-1}f' - (I - B(\gamma')^{-1}f') \\ &\leq N_T(\theta_s(f')_s - f')e^{b(k(T)-k(0))} + N_T((I - B_s(\gamma))^{-1}(I - B_s(\gamma') \\ &\quad - (I - B_s(\gamma)))(I - B_s(\gamma'))^{-1}f') + N_T((I - B_s(\gamma'))^{-1}f' - (I - B(\gamma'))^{-1}f') \\ &\leq N_T(\theta_s(f')_s - f')e^{b(k(T)-k(0))} + N_T(B_s(\gamma') - B_s(\gamma))e^{2b(k(T)-k(0))}N_T(\Pi_s f') \\ &\quad + N_{HT}((I - B_s(\gamma'))^{-1} - (I - B(\gamma'))^{-1}N_T(f') \\ &\leq N_T(\theta_s(f')_s - f')e^{b(k(T)-k(0))} + N_T(B(\gamma') - B(\gamma))e^{2b(k(T)-k(0))}N_T(\Pi_s f') \\ &\quad + N_{HT}((I - B_s(\gamma'))^{-1} - (I - B(\gamma'))^{-1}N_T(f') \end{aligned}$$

For $\varepsilon > 0$ choose a partition t of [-r, T] such that if s is a refinement of t then $N_T(\theta_s(f')_s - f') < \varepsilon/(3e^{b(k(T)-k(0))})$ and by Theorem 3.8, $N_{HT}((I - B_s(\gamma'))^{-1} - (I - B(\gamma'))^{-1}) < \varepsilon/(3(N_T(f) + 1))$. Also choose $\delta > 0$ so that $N_T(B(\gamma') - B(\gamma)) < \varepsilon/(3(N_T(f) + 1))$. Again $N_T(\Pi_s f) \leq N_T(f) + 1$ and so the result follows. □

Since the approximate solution $h_a = h_1(\gamma, s)$ or $h_a = h_2(\gamma, s)$ and h´ are in

$\Pi_s G_H$ it is natural to define the error functional for the parameter estimation algorithm by $J(\gamma) = N_H(h_a - h')^2$, which simplifies to $J(\gamma) = (h_a - h')_s{}^*(K_s)^{-1}(h_a - h')_s$ where $(K_s)^{-1}$ is as defined in Section 2. One should recall that the Hellinger norm dominates the supremun and variational normsand hence minimization of $J(\gamma)$ will tend to minimize the supremum and variational norm of the error $h_a - h'$.

Using vector matrix notation introduced in the previous section, one obtains

$$h_1(\gamma, s)(s_p) = [\theta_s(f')_s](s_p) + [B(\gamma)\theta_s(h')_s](s_p)$$

and $h_1(\gamma, s)$ is uniquely determined by the matrix equation

$$(h_1(\gamma, s))_s = (f')_s + (LB_s)^*(K_s)^{-1}(h')_s$$

Similarly $h^2(\gamma, s)$ is uniquely determined as the solution of the matrix equation

$$(h_2(\gamma, s))_s = (f')_s + M_s(h_2(\gamma, s))_s$$

where the matrix M_s is as defined in Section 4.

Subject to these choices the parameter estimation algorithm can be formulated as an optimization problem in a finite dimensional vector space.

Parameter Estimation Algorithm.

Minimize $J(\gamma) = (h_a(\gamma, s) - h')_s{}^*(K_s)^{-1}(h_a(\gamma, s) - h')_s$

Subject to

1) $(h_a(\gamma, s))_s = (f')_s + (LB_s(\gamma))^*(K_s)^{-1}(h')_s$

or

2) $(h_a(\gamma, s))_s = (f')_s + M_s(\gamma)(h_a(\gamma, s))_s$

A common method of minimizing an error functional is to employ an iterative Newton or conjugate-gradient routine. Such procedures frequently require differentiability of the error functional $J(\gamma)$ with respect to the parameter γ. In the parameter estimation algorithm differentiability of $\Pi_s B(\gamma)\Pi_s$ and $B_s(\gamma)$ with respect to γ is required. Equivalently $(LB_s(\gamma))^*(K_s)^{-1}$ and $M_s(\gamma)$ must be differentiable with respect to γ. For many operators such differentiability conditions are satisfied and optimization can proceed. It will be shown that such differentiability conditions are satisfied for the basic operator $B(\gamma)$ of Example 3.1, which arises in the formulation of delay differential equations.

The choice of which approximation $h_1(\gamma, s)$ or $h_2(\gamma, s)$ to use for the solution of the parameter estimation problem is dependent upon physical and numerical considerations. Recall that the defining equation for the first approximate solution is $h_1(\gamma, s) = \theta_s(f')_s + \Pi_s B(\gamma)_s \theta_s(h')_s$. Assuming f' is known, as is frequently the case, this approximation would be sensitive to inaccuracies in the measurements of the output functions h'. The second approximate solution was defined as the solution of $h_2(\gamma, s) = \theta_s(f')_s + B_s(\gamma)h_2(\gamma, s)$ and is independent of the output measurements. If the output data is sparse then it would appear that the algorithm using the second approximate solution will yield more accurate estimates. Computationally, greater numerical effort is required when the second approximate solution is used. That is, solution of the equation $h = f' + B_s(\gamma)h$ results in greater computation time to complete an iteration in the parameter estimation algorithm. An algorithm is usually easier to implement when the approximation $h_1(\gamma, s)$ is used. For some specific delay differential equations many simplifications have been found for implementing the algorithm, as indicated in the following example.

Example 3.6. To illustrate the significance of these comments, we consider the parameter estimation problem for delay differential equations of the form

$$
\begin{aligned}
h(t) &= f(t) && -r \le t \le 0 \\
[dh/dt](t) &= u(t) + \sum_{i=1}^{m} \alpha_1 h(t - \rho_i) && 0 \le t \le T
\end{aligned}
$$

where u is an integrable control function, the α_i are $d \times d$ matrices, and $0 \le \rho_1 < \rho_2 < \cdots < \rho_m \le r$. The system can be put in the form $h = f + B_1(\gamma)h$ by defining $f(t) = h(0) + \int_{[0,t]} u(s)\, ds$, $\gamma = (\alpha_1, \alpha_2, \cdots, \alpha_m, \rho_1, \rho_2, \cdots, \rho_m)$, and

$$
[B_1(\gamma)h](t) = \begin{cases} 0 & -r \le t \le 0 \\ \int_0^t \sum_{i=1}^{m} \alpha_i h(s - \beta_i) ds & 0 \le t \le T \end{cases}
$$

One may show that $B_1(\gamma)$ is in $\boldsymbol{B}$ and thus the theory of the preceding sections concerning the representation and approximation of $B_1(\gamma)$ applies. However, an alternative form of the system is more useful for computational purposes. Let α be a $d \times d$ matrix and $0 \le \rho \le r$ and define the operator $B(\alpha, \rho)$ by

$$[B(\alpha, \rho)h](t) = \begin{cases} 0 & -r \le t \le 0 \\ \alpha\int_0^t h(\tau - \rho)d\tau & 0 \le t \le T \end{cases}$$

Then $B(\alpha, \rho)$ is in $\boldsymbol{B}$ and $B_1(\gamma) = \Sigma_{i=1,m} B(\alpha_i, \rho_i)$. The representation and approximation of $B_1(\gamma)$ can be formed as the sum of representations and approximations of the $B(\alpha_i, \rho_i)$ for $i = 1, \cdots, m$. A wide class of problems can now be treated using only the parameterized representation of $B(\alpha, \rho)$.

To obtain this matrix representation, let $k(t) = 1 + r + t$ and $s = \{-r + pc\}_{p=0,n}$ with $c > 0$ be a partition of $[-r, T]$ satisfying $r = p'c$ for some positive integer p' then

$$[\Pi_s B(\gamma)\Pi_s h](s_p) = [B(\gamma)\Pi_s h](s_p) = [(LB(\gamma)_s)^*(K_s)^{-1}\Pi_s h](p)$$

Here, see page 94,

$$(LB(\gamma)_s)^*(K_s)^{-1}(p, 0) = LB(\gamma)(s_0, s_p)^*(1 + 1/c) + LB(\gamma)(s_1, s_p)^*(-1/c)$$

and

$$(LB(\gamma)_s\ )^*(K_s)^{-1}(p, q) = LB(\gamma)(s_{q-1}, s_p)^*(-1/c) + LB(\gamma)(s_q, s_p)^*(2/c)$$
$$+ LB(\gamma)(s_{q+1}, s_p)^*(-1/c)$$

Let $s_{q'} \le -\rho < s_{q'+1}$. Then $s_{p-p'+q'} \le s_p - \rho < s_{p-p'+q'+1}$ for $s_p > 0$ and one obtains using the representation of $LB(\gamma)$ calculated in Example 3.1

$(LB(\gamma)_s)^*(K_s)^{-1}(p, q) =$

$$\begin{cases} 0I_d & p = p' + 1 \quad q < q' \\ \alpha(s_{q'} + c + \rho)^2/2c & q = q' \\ \alpha(c^2 - 2(s_{q'} + \rho)c - 2(s_{q'} + \rho)^2)/2c & q = q' + 1 \\ \alpha(s_{q'} + \rho)^2/2c & q = q' + 2 \\ 0 & q > q' + 2 \end{cases}$$

$$\begin{cases} 0 & p = p' + 2 \quad q < q' \\ \alpha(s_{q'} + c + \rho)^2/2c & q = q' \\ \alpha(2c^2 - (s_{q'} + \rho)^2)/2c & q = q' + 1 \\ \alpha(2c^2 - (s_{q'} + \rho + c)^2)/2c & q = q' + 2 \\ \alpha(s_{q'} + \rho)^2/2c & q = q' + 3 \\ 0 & q > q' + 3 \end{cases}$$

$$\begin{cases} 0 & p > p' + 2 \quad q < q' \\ \alpha(s_{q'} + c + \rho)^2/2c & q = q' \\ \alpha(2c^2 - (s_{q'} + \rho)^2)/2c & q = q' + 1 \\ \alpha c & q' + 1 < q < q' + (p - p') \\ \alpha(2c^2 - (s_{q'} + \rho + c)^2)/2c & q = q' + (p - p') \\ \alpha(s_{q'} + \rho)^2/2c & q = q' + (p - p' + 1) \\ 0 & q > q' + (p - p' + 1) \end{cases}$$

Clearly $(LB(\gamma)_s)^*(K_s)^{-1}(p, q)$ is differentiable with respect to the components of α. Differentiability with respect to ρ also clearly holds when $s_q < \rho < s_{q+1}$. If $\rho = s_{q'}$ one obtains

$(\partial/\partial\rho)(LB(\gamma)_s)^*(K_s)^{-1}(p, q) =$

$$\begin{cases} 0 & p = p' + 1 \quad q < q' \\ \alpha & q = q' \\ -\alpha & q = q' + 1 \\ 0 & q > q' + 1 \end{cases}$$

$$\begin{cases} 0 & p = p' + 2 \quad q < q' \\ \alpha & q = q' \\ 0 & q = q' + 1 \\ -\alpha & q = q' + 2 \\ 0 & q \geq q' + 3 \end{cases}$$

$$\begin{cases} 0 & p > p' + 2 \quad q < q' \\ \alpha & q = q' \\ 0 & q' + 1 \leq q \leq q' + (p - p') \\ -\alpha & q = q' + (p - p') \\ 0 & q \geq q' + (p - p' + 1) \end{cases}$$

Using these calculations one may implement the parameter estimation algorithm to estimate coefficient and delay parameters appearing in a delay differential equation. Numerical results of such an implementation appear in [10, 11].

7. Numerical Control Problems

In this section, we show that the discretization methods developed in this chapter provide a flexible means of obtaining approximate solutions to the optimal control problems considered in Chapter II. Given B in ***B*** and $A = (I - B)^{-1}$, we obtain, based upon the analysis of Chapters II and III, approximate solutions to the following optimal control problems:

I. Minimize: $J(u, h) = (1/2)c_1 N_H(Au)^2 + (1/2)N_H(Ch)^2$
Subject to: $h = f + Bh + u$
$h(t) = 0 \qquad T - r \leq t \leq T$

and for finite dimensional state space systems

II. Minimize: $J(u, h) = (1/2)N_H(u)^2 + (1/2)N_H(Ch)^2$
Subject to: $h = f + Bh + u$
$h(T) = 0$

In these problems f is to be thought of as a known constantly acting disturbance and c_1 in problem I is a weighting parameter.

Three examples are considered. In the first example numerical approximations are obtained for the solution of optimal control problem I and for the feedback operator appearing in the solution of this problem. In the second example the response obtained from optimization problem II is computed using two equivalent representations for the response. The objective of the third example is to provide a comparison of the results obtained from use of the two alternative cost functionals to determine a control which steers a system response to zero at a terminal time.

Approximate solutions to a system $h = f + Bh$ may be obtained by the Galerkin method $\Pi_t h = \Pi_t f + \Pi_t B \Pi_t h$, with corresponding matrix equation $h_t = f_t + (LB_t)^*(K_t)^{-1}h_t$, or by the modified method of Corollary 3.10 $\Pi_t h = \Pi_t f + B_t \Pi_t h$, with corresponding matrix equation $h_t = f_t + M_t h_t$.

<u>Example</u> 3.7. Consider the system $h = f + Bh + u$ with

$$[Bh](t) = \begin{cases} 0 & -r \leq t \leq 0 \\ d\int_0^t (t - s)h(s)ds + \beta\int_0^t h(s - r) + \gamma\int_0^t h(s)ds & 0 \leq t \end{cases}$$

The control objective is to steer the solution to zero over a terminal interval $[T - r, T]$. Using the methods and notation of section 3 in Chapter II with $c_1 = 1$ for the solution of optimal control problem I, the solution can be written as

$$h_0 = P_{T-r}(I - C^2)^{-1}\{Af + k_1\lambda_0\}$$

and the corresponding control as

$$u_0 = A^{-1}C^2h - A^{-1}(I - C^2)(I - P_{T-r})(I - C^2)^{-1}Af + A^{-1}(I - C^2)P_{T-r}(I - C^2)^{-1}k_1\lambda_0$$

For f in G_H, $P_{T-r}f$ and $(I - P_{T-r})f$ may be computed using numerical quadrature rules, and it remains to approximate $h = Af = (I - B)^{-1}f$, i.e., solutions of $h = f + Bh$.

For illustration, let $\alpha = -2$, $\beta = 1$, $\gamma = 1$, $f(t) = 1$ and $k(t) = 2 + t$ for $-1 \le t \le 3$. Let $t_p = -1 + p/128$ for $p = 0, 1, \cdots, 128$. Using the Galerkin method, the approximate solution of the system with and without control are depicted in Figure 3.4.

In this example the control contains a feedback term $w = A^{-1}C^2h = (I - B)C^2h$. Although Theorem 3.2 guarantees $\Pi_t A^{-1}C^2\Pi_t$ converges to $A^{-1}C^2$, the operator

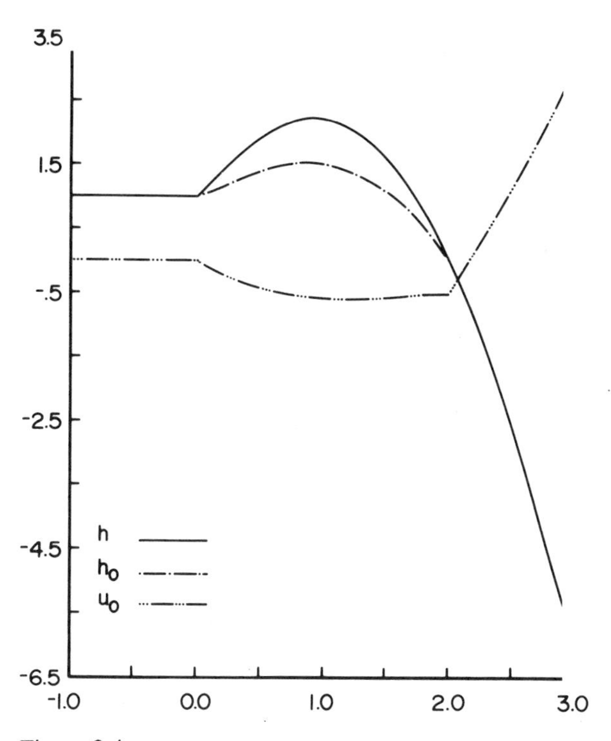

Figure 3.4

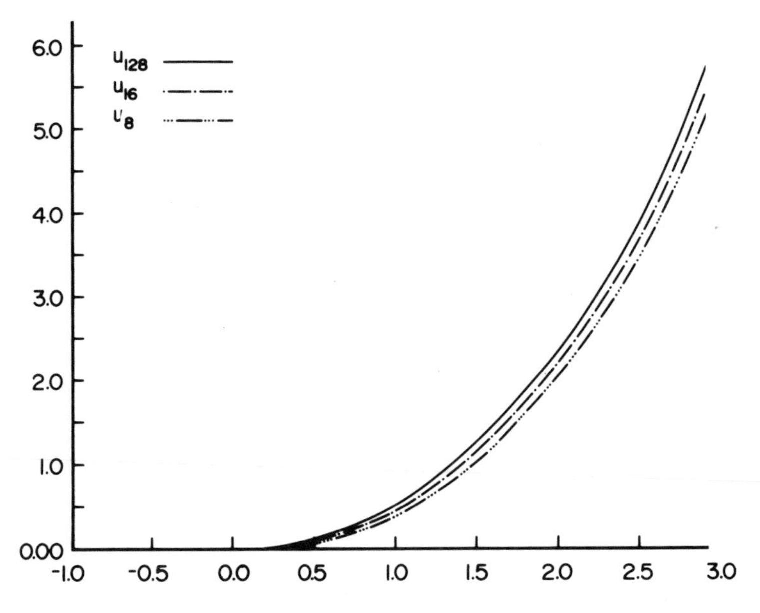

Figure 3.5

$\Pi_t A^{-1}C^2\Pi_t$ is not causal. Digital implementation of such a control law requires an alternate approximation. Let $F = A^{-1}C^2$ and $t = \{t_p\}_{p=0,n}$ be a partition of [-1, 3] with tp = -1 + pc. Then F_t as defined in equation 3.10 is causal and converges to F by Theorem 3.7. The convergence of $w_c = F_s h_0$ with h_0 as computed above and c = 1/8, 1/16, 1/120 is depicted in Figure 3.5. We have also calculated max(| $w_{128}(t_p)$ - $w_{256}(t_p)$ |) = .022475.

Example 3.8. A control which steers the response of the finite dimensional state space system h = f + Bh + u with $[Bh](t) = \alpha\int_{0,t} h(\tau)d\tau$ for $t > 0$ and α a $d \times d$ matrix to zero at a terminal time T can be obtained using optimal control problem II. As derived in Chapter II the optimal response and control are given by

$$h = A_1 f + k_1\lambda_0$$
$$u = (I - B_1)^{-1}C^2 h + k_2\lambda_0$$

where

$$A_1 = (I - B - (I - B_1)^{-1}C^2)^{-1}$$

$$B_1h = B^*h - K(\ ,T)[B^*h](0)$$

$$k_2 = (I - P_0)(I - B_1)^{-1}K(\ ,T)$$

$$k_1 = A_1k_2$$

$$\lambda_0 = -[A_1f](T)/k_1(T)$$

At first glance computation of h and u seems somewhat formidable. However, as shown in Chapter II, the following simplifications can be made:

$$[C^2f](t) = \int_0^t (t - s)f(s)\,ds$$

$$[B_1f](t) = \alpha^*tf(0) - \alpha^*\int_0^t f(\tau)\,d\tau$$

$$[(I - B_1)^{-1}f](t) = f(t) + (1 - e^{-\alpha^*t})f(0) - \alpha^*\int_0^t e^{-\alpha^*(t-\tau)})f(\tau)\,d\tau$$

and if α is invertible

$$[(I - B_1)^{-1}C^2f](t) = (\alpha^*)^{-1}\{\int_0^t f(\tau)\,d\tau - \int_0^t e^{-\alpha^*(t-\tau)}f(\tau)\,d\tau\}$$

Thus computation of the optimal response and control essentially reduces to the problem of calculating A_1f with f in G_H. In the scalar case, one obtains using Laplace transforms

$$[A_1f](t) = f(t) + \alpha\int_0^t\cosh(\sqrt{\alpha^2 + 1}\ (t - \tau))f(\tau)d\tau$$
$$+ \sqrt{\alpha^2 + 1}\int_0^t\sinh(\sqrt{\alpha^2 + 1}\ (t - \tau))f(\tau)d\tau$$

In the vector case, one needs to approximate the solution of $h = f + Bh + (I - B_1)^{-1}\cdot C^2h + k_2\lambda_0$. Due to the simple formulas for B, $(I - B_1)^{-1}$ and C^2, calculation of

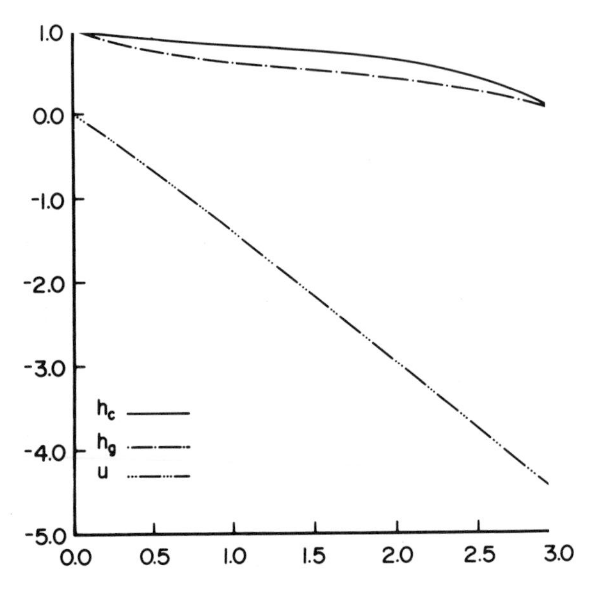

Figure 3.6

approximate solutions to this system by our discretization methods is straightforward. The following modified Galerkin method

$$\Pi_t h = \Pi_t f + \Pi_t B \Pi_t h + \Pi_t (I - B_1)^{-1} \Pi_t \Pi_t C^2 \Pi_t h$$

and the associated matrix equation

$$h_t = f_t + \{(LB_t)^*(K_t)^{-1} + (L(I - B_1)^{-1}{}_t)^*(K_t)^{-1}(LC^2{}_t)^*(K_t)^{-1}\}h_t$$

are immediately suggested.

For numerical illustration let $T = 3$, $\alpha = 1$, $f(t) = 1$ and $k(t) = 1 + t$ for $0 \leq t \leq 3$. In Figure 3.6 the optimal response obtained by using both the closed form formula, hc, and the Galerkin approximation, hg, is depicted. We have also graphed the optimal control.

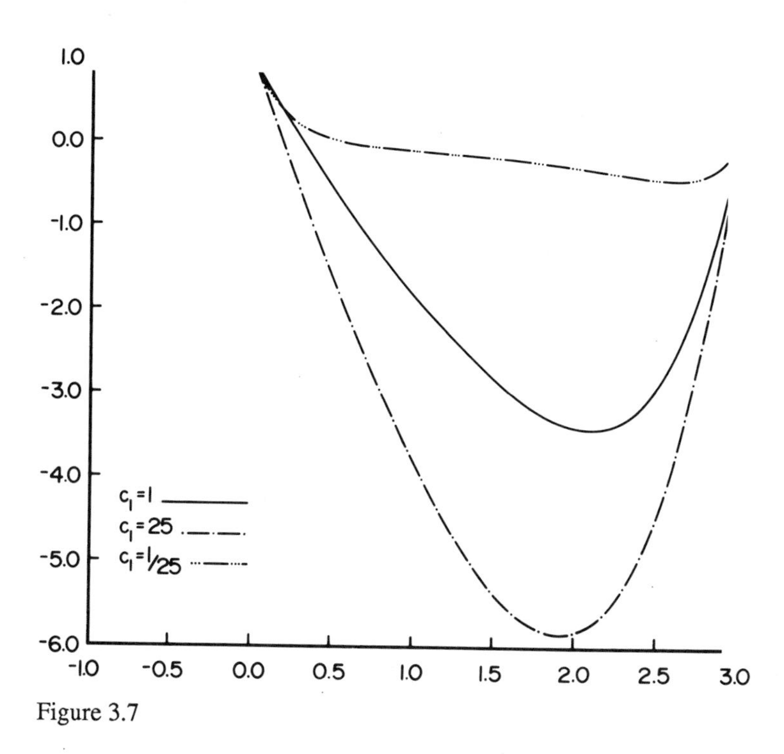

Figure 3.7

Example 3.9. Optimal control problem I can be used to achieve the same control objective as considered in Example 3.8. In this case the optimal response and control are given by

$$h = (I - C^2/c_1)^{-1}\{Af + k_1\lambda_0/c_1\}$$
$$u = (1/c_1)A^{-1}C^2h + A^{-1}k_1\lambda_0/c_1$$

Derivation of these formulas requires a slight modification of the argument given in Chapter II. Dependence of the solution of this optimal control problem upon the weighting parameter c_1 is depicted in Figures 3.7 and 3.8. These Figures should be compared with Figure 3.6 for an indication of the differences which result from use of the alternate cost functionals.

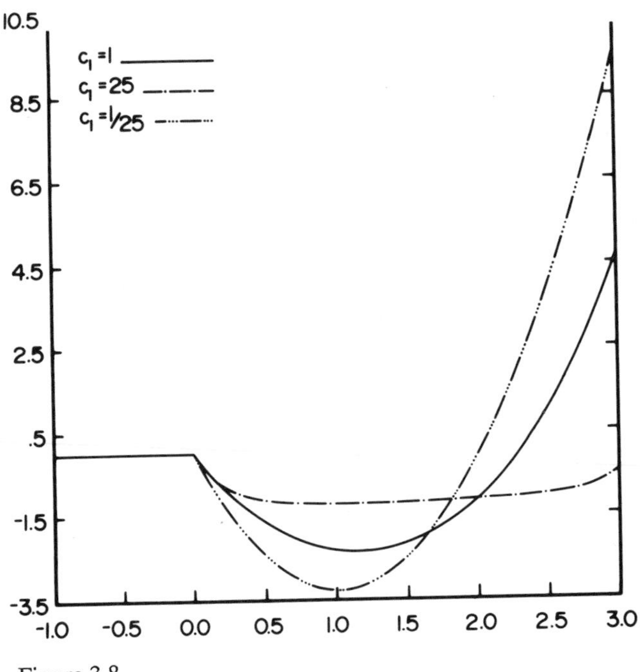

Figure 3.8

Summary

Many problems in numerical analysis have been approached within a reproducing kernel Hilbert space setting. Golomb and Weinberger [20] show that problems of numerical differentiation, interpolation, quadrature, Fourier analysis and synthesis, summation of series, and the solution of differential and integral equaions reduce to the problem of approximating a linear functional of an unknown function given values of a finite set of other linear functionals. In their work, reproducing kernels are used to obtain representors of the given linear functionals and to obtain formulas for best approximations and error bounds.

Finite basis or projection methods (spectral theory, Galerkin, and collocation methods) are frequently used to obtain operator approximations and approximate solutions to operator equations [19, 21]. Within a reproducing kernel Hilbert space the K-polygonal functions provide finite dimensional subspaces upon which to base projection methods. The text [24] by R. E. Moore contains a discussion of the approximation of linear operators and functionals within a reproducing kernel Hilbert space setting. In particular Galerkin and collocation methods are discussed.

DeBoor and Lynch [16] relate the theory of splines to best approximation methods in reproducing kernel Hilbert spaces and develop a class of generalized splines determined by an n-th order differential operator and a collection of linear functionals. They show that these generalized splines retain familiar minimum properties of splines.

In this chapter the use of projections to obtain the approximation $\Pi_t F \Pi_t$ to an operator F in ***A*** or ***B*** is standard. However, for us the operator norm N_{HT}, first introduced in our paper [29], is essential for the analysis of convergence of such operator approximations. Clearly $N_{HT}(F) \leq N_T(F)$ and convergence of operator approximations cannot be with respect to the norm N_T since operators in ***A*** need not be compact.

The standard Galerkin approximate solution to the operator equation $h = f + Bh$ is defined $\Pi_t h = \Pi_t f + \Pi_t B \Pi_t h$ with t a partition of the interval [-r, T]. The operator approximation B_t, defined in Section 3, is in the class ***B*** and, consequently, is a causal operator. The approximation $\Pi_t B \Pi_t$ need not be causal.

Such approximations arise naturally in the analysis of feedback control systems [30] and the fact that B_t belongs to ***B*** allows a unified analysis. Consider a system diagrammed as shown in Figure 3.9 with A in ***A*** and the feedback operator D in ***B***. The digital implementation of such a control scheme requires an approximation to the feedback operator D. The approximation $u = \Pi_t D \Pi_t h$ is inappropriate since present values of the control u depend on future values of h, i.e., if $t_{i-1} \leq t < t_i$ then u(t) depends on $h(t_i)$. Such considerations motivated the approximation D_t and consideration of the equation $h = f + Bh + D_t h$.

Reproducing kernel Hilbert space methods have been used by G. Wahba [34] to obtain approximate solutions to two point boundary value problems for a class of linear operator equations and by Ciarlet and Varga [15] to obtain approximations to the Green's function associated with such boundary value problems. The classes of equations considered by these authors fall within our general linear hereditary system framework.

Although analysis of the Galerkin approximate solutions $\Pi_t h = \Pi_t f + \Pi_t B \Pi_t h$ has not been included in this chapter, numerical experiments indicates the

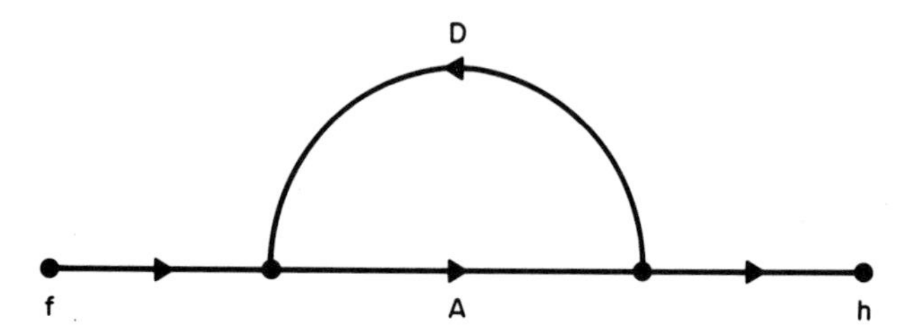

Figure 3.9

convergence of such approximations when B belongs to ***B***. Using our discretization techniques one can show that the solution of $h_a = \Pi_t f + \Pi_t B \Pi_t h_a$ is of the form $h_a(t_i) = f(t_i) + \Sigma_{j=0,i} w(i, j) h_a(t_j)$. Weighted sums of this type arise in the approximate solution of Volterra integral equations when numerical integration rules are used to approximate the integrals [23]. Such approximations may be compared with numerical methods for the solution of delay differential equations [17, 25].

Recent research by Banks and Burns [4], Rosen [31], Kunish [22], and Gibson [18] has provided a unified approach to the approximation of solutions to delay differential equations, functional differential equations, hereditary systems, and related system problems. The infinite dimension nature of such systems leads to a description as an abstract evolution equation and to approximations by semi group methods. Approximations by finite sets of ordinary differential equations [1, 3, 6, 13] and by difference equations [8, 28, 33] have been obtained.

The system identification problem has been defined as the determination of a system operator from input-output or input-response data. Such problems have also been referred to as data interpolation problems by W. A. Porter [27]. Here properties of reproducing kernel Hilbert spaces allow the description of the structure of system operators, a structure which is simplified in the time invariant case. Our results show that if the system can be "pulsed" with appropriate inputs and the corresponding responses are measured exactly then the system operator can be identified.

In system modeling and input-output description is appropriate when little *a priori* knowledge of the system dynamics is available. If system dynamics are known then a set of differential or integral equations may be used as the model and the identification problem reduces to the determination of parameters appearing in the model equations, the parameter estimation problem.

Our ideas on system identification and parameter estimation originated in the papers [10, 11, 29] and the thesis [9] by S. B. Benz. Parameter estimation methods have been introduced for delay systems by Banks, Burns, and Cliff [5], Banks [2], Banks and Lamm [7], Burns and Hirsh [14], Rosen [32], and Pearson [26]. Reproducing kernel Hilbert space methods are used by Zyla and deFigueiredo [36] in the identification of a nonlinear dynamical system. Consideration of system noise or measurement errors leads to the analysis of statistical signal processing problems. The collection of papers [35] edited by H. L. Weinert provides an introduction to the application of reproducing kernel Hilbert space methods to a broad class of detection, estimation, and approximation problems.

References

1. H. T. Banks, Approximation of nonlinear functional differential equation control systems, J. Optim. Theory Appl. 29(1979), 383-408.

2. H. T. Banks, Identification of nonlinear delay systems using spline methods, Proc. International Conference on Nonlinear Phenomena in the Mathematical Sciences, Univ. Texas, Arlington, Academic Press, New York, 1980)

3. H. T. Banks and J. G. Rosen Spline approximations for linear nonautonomous delay systems, J. Math. Anal. and Appl., 96(1983), 226-268.

4. H. T. Banks and J. A. Burns, Hereditary control problems: numerical methods based on averaging approximations, SIAM J. Control Opt. 16(1978), 169-208.

5. H. T. Banks, J. A. Burns, and E. M. Cliff, Parameter estimation and identification for systems with delays, SIAM J. Control Opt. 19(1981), 791-828.

6. H. T. Banks and F. Kappel, Spline approximations for functional differential equations, J. Differential Eqs. 34(1979), 496-522.

7. H. T. Banks and P. K. Daniel Lamm, Estimation of delay and other parameters in nonlinear functional differential equations, SIAM J Control Opt. 21(1983), 895-915.

8. H. T. Banks, I. G. Rosen, and K. Ito, A spline based technique for computing Riccati operators and feedback controls in regulator problems for delay equations, SIAM J. Sci. Stat. Comput. 5(1984), 830-855.

9. S. L. Benz, Parameter Estimation in a Reproducing Kernel Hilbert Space for Linear Hereditary Systems, Ph.D. Dissertation, Clemson University, 1981.

10. S. L. Benz and R. E. Fennell, Delay differential equations - approximate solutions and parameter dependence, Proceedings of the Fifteenth Southeastern Symposium on Systems Theory, March 1983.

11. S. L. Benz, R. E. Fennell, and J. A. Reneke, Hereditary systems: approximate solutions and parameter estimations, Applicable Analysis 17(1984), 135-155.

12. J. M. Bownds, On an initial-value method for quickly solving Volterra integral equations: a review, J. Opt. Theory Appl. 24(1978), 133-151.

13. J. A. Burns and E. M. Cliff, Methods for approximating solutions to linear hereditary quadratic optimal control problems, IEEE Trans. Automatic Control, 23(1978), 21-36.

14. J. A. Burns and P. D. Hirsch, A difference equation approach to parameter estimation for differential-delay equations, Appl. Math. Comp. 7(1980), 281-311.

15. P. G. Ciarlet and R. S. Varga, Discrete Variational Green's Functions II, Numer. Math, 16(1970), 115-128.

16. C. DeBoor and R. E. Lynch, On splines and their minimum properties, J. Math. Mech. 15(1966), 953-969.

17. A. Feldstein, Higher order methods for state-dependent delay differential equations with nonsmooth solutions, SIAM J. Numer. Anal. 21(1984), 844-863.

18. J. S. Gibson, Linear quadratic optimal control of hereditary systems: infinite dimensional Riccati equations and numerical approximations, SIAM J. Control Opt. 21(1983), 95-139.

19. I. C. Gohberg, and I. A. Feldman, Convolution Equations and Projection Methods for their Solution, Translations of Mathematical Monographs Vol. 14, AMS, Providence, R.I., 1974.

20. M. Golomb and H. F. Weinberger, Optimal approximation and error bounds, in On Numerical Approximation - edited by R. E. Langer, Univ. of Wisconsin Press, 1959.

21. M. A. Krasnosel'skii, G. M. Vainikko, P. P. Zabreiko, YaB Rutitskii, and V. Ya. Stetsenko, Approximate Solutions of Operator Equations, Wolters-Noordroff Publishing, Groningin, 1972.

22. K. Kunisch, Approximation schemes for linear quadratic optimal control problems associated with delay equations, SIAM J Control Opt. 20(1982), 506-540.

23. P. Linz, Analytical and Numerical Methods for Volterra Equations, SIAM Studies in Applied Mathematics, SIAM Philadelphia, 1985.

24. R. E. Moore, Computational Functional Analysis, Ellis Horwood Limited, Chichester, England, 1985.

25. K. W. Neves, Automatic integration of functional differential equations: An approach, ACM Trans. Math. Software, 1(1975),

26. A. E. Pearson, Decoupled delay estimation in the identification of differential delay systems, Automatica 20(1984), 761.

27. W. A. Porter, Data interpolation, causality structure, and system identification. Inform. and Control 29(1975), 217-233.

28. D. C. Reber, Finite difference techniques for solving optimization problems governed by linear functional differential equations, J. Differential Eqs. 32(1979), 193-232.

29. J. A. Reneke, and R. E. Fennell, An identification problem for hereditary systems, Applicable Analysis 11(1981), 167-183.

30. J. A. Reneke and R. E. Fennell, RKH space approximations for the feedback operator in a linear hereditary control system, submitted for publication.

31. I. G. Rosen, A discrete approximation framework for hereditary systems, J. Differential Eqs 40(1981), 377-449.

32. I. G. Rosen, Discrete approximation methods for parameter identification in delay systems, SIAM J Control Opt. 22(1984), 95-120.

33. I. G. Rosen, Difference equation state approximations for nonlinear hereditary control problems, SIAM J. Cont. Opt. 22(1984), 302-326.

34. G. Wahba, A class of approximate solutions to linear operator equations, J. Approx. Theory 9(1973), 61-77.

35. H. L. Weinert, Reproducing Kernel Hilbert Spaces, Applications in Statistical Signal Processing, Hutchinson Ross Publishing Company, Stroudsburg, PA, 1982.

36. L. V. Zyla, and deFigueiredo, Nonlinear system identification based on a Fock space framework, SIAM J. Cont. Opt. 21(1983), 931- 939.

IV

Optimal Control of Stochastic Hereditary Systems

1. Introduction to Stochastic Systems

In this chapter, the study of hereditary systems is continued by considering systems which contain a stochastic, or random, element. Many of the results obtained in previous chapters will hold for stochastic systems, with proof of the results only requiring the appropriate choice of a base space X. However, probabilistic considerations will endow the stochastic results with interpretations and properties which are very different from the corresponding deterministic results. Before outlining the contents of this chapter, we give some motivation for the study of stochastic systems.

Motivational remarks. Stochastic processes are often used to explicitly indicate imprecision in a model. A stochastic process may represent an input which cannot be precisely measured, or an input which is unpredictable beyond an estimate of its mean and mean square values. For example, the velocity of a wind is unpredictable, although a forecast of 5 to 10 mph may give the correct range of values. A measurement of the wind velocity may be inaccurate, with the error well modeled by a stochastic process.

Along with the above issues of modeling completeness, stochastic models may be useful because of the different quantitative and qualitative properties that solutions of random equations possess. For instance, consider the deterministic competing species model $X' = X(1 - X - Y)$, $Y' = Y(X - c)$. For appropriate initial conditions, the predator (Y) becomes extinct with certainty if $c = 1.01$, but survives with certainty if $c = .99$. This discontinuity in the probability of extinction does not occur for analogous stochastic models. If c is random, or if random disturbances are added to the equations for X' and Y', values of the parameters may be found so that the probability of extinction assumes any value between 0 and 1. In addition, a comparison of input and output variances may provide an easy form of sensitivity analysis. Thus, a stochastic model may be richer than its deterministic counterpart.

Chapter outline. Our appoach to stochastic systems is unsophisticated in that we require little of the heavy machinery (see Wong and Hajek [27] for an exposition of martingale theory) developed specifically for stochastic systems. We consider this an advantage, as it enhances the mathematical and intuitive accessibility of our work.

Starting from basic principles of measure theory, we develop in Section 2 the terminology necessary to discuss stochastic hereditary systems. In Section 3, time series representations of stochastic processes are presented, and in Section 4 we give a brief discussion of stochastic differential equations. With the background provided in these sections, our analysis proceeds essentially the same as for deterministic systems.

In the spirit of Chapter II, hereditary systems are analyzed within the structure of the RKH space generated by a Hellinger integral, and the operator spaces $\boldsymbol{A}$ and $\boldsymbol{B}$. The relevant definitions and preliminary results are given in Section 2, and optimal control problems for state space and general hereditary systems are solved in Section 3. A calculation of the output covariance matrix, based on a time series method of Parzen (see [21] or [25]), is also given in Section 3.

The important case of systems with state dependent noise is not covered by the $\boldsymbol{A}/\boldsymbol{B}$ description of hereditary systems. To include such systems we must extend the class of operators considered. The extension, given in Section 4, is strongly motivated by analytical and modeling considerations. A control problem for extended hereditary systems is solved in Section 4.

2. Background for Stochastic Hereditary Systems

In this section, we give the definitions which will be required to apply the results derived in previous chapters. We also introduce some of the concepts which are unique to probability spaces and stochastic processes.

Spaces of stochastic processes. The underlying probability space is denoted $(\Omega, \boldsymbol{F}, P)$. That is, Ω is a set (typically unknown in applications) and $\boldsymbol{F}$ is a σ-algebra of subsets of Ω. For a fixed ω in Ω, all functions and parameters are deterministic, so Ω characterizes the randomness in the system. The set function P is a probability measure on $(\Omega, \boldsymbol{F})$, so that $P(\Omega) = 1$.

An $\boldsymbol{F}$-measureable scalar function $x : \Omega \to R$, called a random variable, is in $L_2(\Omega)$ if $\int_\Omega x^2(\omega)\, dP(\omega) < \infty$. The space $X = L_{2,d}(\Omega)$ consists of all vector-valued functions $x : \Omega \to R^d$ such that each component of x is in $L_2(\Omega)$. The inner product of x and y (having components x_i and y_i, respectively) in X is given by $<x, y> = \int_\Omega x_1y_1 + \cdots + x_dy_d\, dP$. Here, and throughout the chapter Ω-dependence is suppressed in the notation. The corresponding norm satisfies $|x|^2 = <x, x>$.

It is often helpful to use statistical terminology instead of inner product notation. The expected value of x in X is $E(x) = \int_\Omega x\, dP$. We will usually assume $E(x) = 0$, the mean zero case. The covariance matrix of x is $E(xx^*)$, where x^* is the transpose of the column vector x.

The function space G consists of all right continuous functions $f : S \to X$. In this chapter, we restrict our attention to the finite time interval $S = [-r, T]$. Elements of G are second order stochastic processes. Depending on context, it is convenient to view f in G in alternate ways. For a fixed t in S, f(t) is a function from Ω into R^d so f may be thought of as a function $f(t, \omega)$ from $S \times \Omega$ into R^d with f(t) being shorthand for $f(t, \cdot)$. For applications, the "sample path" characterization of f is intuitively appealing. The function $f(\cdot, \omega)$, with ω in Ω fixed, from S into R^d is a sample path or "realization" of the process f and may represent a particular outcome of an experiment. The expected value of f is the (weighted) average of all sample paths.

Two subspaces of G are important in the sections to follow. We define G_0 to be the set of all f in G for which $P_0f = 0$ and $E(f(t)) = 0$ for each t in S. The space G_z consists of all z in G_0 such that z has independent increments: if $-r \leq s \leq t \leq u \leq v \leq T$ and $1 \leq i, j \leq d$, $E[\{z(v) - z(u)\}_i\{z(t) - z(s)\}_j] = E[\{z(v) - z(u)\}_i]E[\{z(t) - z(s)\}_j]$.

Figure 4.1 illustrates the independent increment property of an element z of G_z, with dimension $d = 1$. Two sample paths are shown for $0 \leq t \leq s$, with $z(s, \omega_1) = z(s, \omega_2) = A$. Consider the problem of predicting the change in z over a small period of time; that is, estimating $z(s + \Delta s, \omega_i) - z(s, \omega_i)$ for $i = 1$ and $i = 2$. If

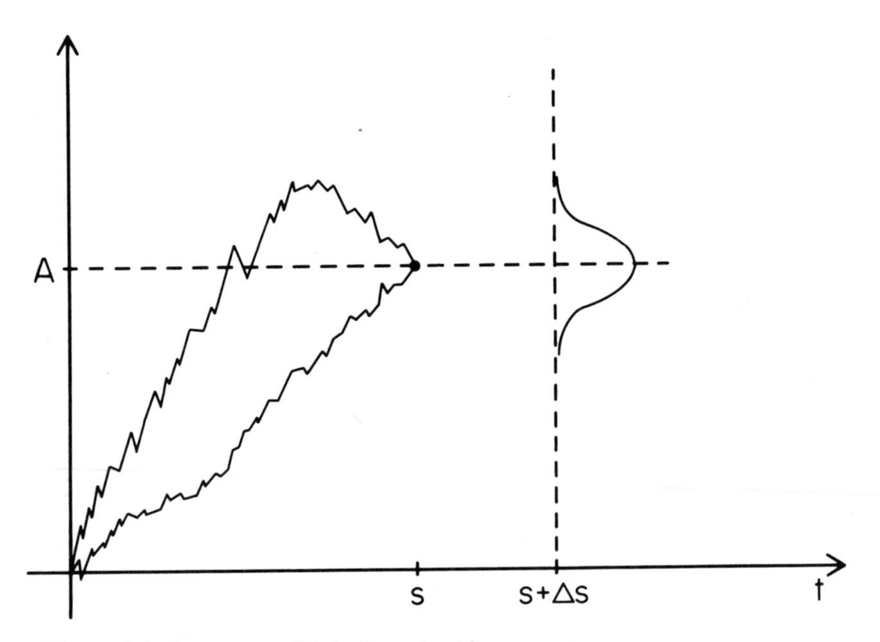

Figure 4.1. A process with independent increments

z is a standard Wiener process (discussed below), then $z(s + \Delta s) - z(s)$ is a normally distributed random variable with mean 0 and variance Δs (normal random variables have bell shaped probability density functions; "ideal" test grades are normally distributed with mean 75). If $z(s) = A$ then $z(s + \Delta s) - z(s)$ is normal with mean 0 and variance Δs, as graphed on the $t = s + \Delta s$, $z = A$ axes. The main implication of independent increments is that our predictions for $z(s + \Delta s, \omega_1)$ and $z(s + \Delta s, \omega_2)$ should be the same, regardless of the differing trends at $t = s$ and the actual value of A. Intuitively speaking, future changes in the process are independent of the past.

Associated with a fixed z in G_z is a collection of σ-algebras $\{\boldsymbol{F}_t : t \text{ in } S\}$ with the properties $\boldsymbol{F}_s \subseteq \boldsymbol{F}_t \subseteq \boldsymbol{F}$ whenever $-r \leq s \leq t \leq T$, and $\sigma(z(s) : -r \leq s \leq t) \subseteq \boldsymbol{F}_t$ for each t in S, where $\sigma(z(s) : -r \leq s \leq t)$ is the smallest σ-algebra of subsets of Ω with respect to which each random variable in $\{z(s) : -r \leq s \leq t\}$ is measurable. The σ-algebra $\boldsymbol{F}_t$ is thought of as containing all information available to an observer at time t.

Exercise. Suppose that z belongs to G_z, $\boldsymbol{F}_t = \sigma(z(s) : -r \leq s \leq t)$, and $\boldsymbol{F} = \boldsymbol{F}_T$. Show that $\{\boldsymbol{F}_t : t \text{ in } S\}$ satisfies the "filtration" properties described above. Show that $\boldsymbol{F}_t = \{\emptyset, \Omega\}$ for $-r \leq t \leq 0$.

A stochastic process is said to be nonanticipating if f(t) is $\boldsymbol{F}_t$-measurable for each t in S; the terminology is meant to indicate that f(t) does not depend on future values of the noise process. Nonanticipating processes can be obtained by taking conditional expectations. The conditional expectation of a random variable x given a σ-algebra $\boldsymbol{F}_t$, denoted $E(x \mid \boldsymbol{F}_t)$, is an $\boldsymbol{F}_t$-measurable random variable satisfying $\int_A x \, dP = \int_A E(x \mid \boldsymbol{F}_t) \, dP$ for each A in $\boldsymbol{F}_t$.

Exercise. Suppose f in G is nonanticipating, $\{t_p\}_{p=0,n}$ is an increasing sequence of numbers with $t_0 = 0$ and $t_n = T$, $\{c_p\}_{p=0,n}$ is a linearly independent set with c_p in X and $E(c_p) = 0$ for each p. Let M be a continuous linear transformation of X such that $M[f(t_p)] = c_p$ for each p. Let

$$g(t) = \begin{cases} 0 & -r \leq t \leq 0 \\ (t-t_{p-1})M[f(t_{p-1})] + g(t_{p-1}) & t_{p-1} \leq t \leq t_p \end{cases}$$

Show that $E(g(t)) = 0$. Under what conditions on $\{c_p\}_{p=0,n}$ is g nonanticipating?

Exercise. If $z \in G_z$ and $\boldsymbol{F}_s = \sigma(z(u) : -r \leq u \leq s)$, the independent increments property of G_z also implies $E(z(t) - z(s) \mid \boldsymbol{F}_s) = 0$ whenever $t \geq s$. Show that if $t \geq s$, then $E(z(t) \mid \boldsymbol{F}_s) = z(s)$. That is, show that z is a $\boldsymbol{F}_s$-martingale.

Exercise. Referring to Figure 4.1, describe the random variable $E(z(s+\Delta s) \mid \boldsymbol{F}_s)$ where $\Delta s > 0$ and $\boldsymbol{F}_s = \{\omega \in \Omega : z(s, \omega) = A\}$.

Stochastic hereditary systems. The systems discussed in this chapter correspond to Example 2 of hereditary systems presented in Chapter 1. For clarity, we redefine the components of a stochastic hereditary system below.

For t in S and f in G, the pseudonorm N_t is given by $N_t(f) = \sup\{ \mid f(x) \mid : -r \leq x \leq t\}$ and the projection P_t satisfies

$$[P_t f](x) = \begin{cases} f(x) & -r \le x \le t \\ f(t) & t \le x \le T \end{cases}$$

In general, we consider k to be a fixed but arbitrary right continuous increasing function on S with $k(-r) = 1$. In the state space case with $r = 0$, k is chosen to be $k(t) = 1 + t$.

The operator space $\boldsymbol{B}$ contains all functions $B : G \to G$ satisfying the following properties: B is linear, $P_0 B = 0$, $EBf = BEf$ for each f in G, and there exists a constant $c = c(B)$ such that whenever f is in G and $-r \le u \le v \le T$,

$$| [Bf](v) - [Bf](u) | \le c \int_u^v N_t(f)\, dk(t)$$

The operator space $\boldsymbol{A}$ contains all operators I - B for B in $\boldsymbol{B}$, and Theorems 1.1 - 1.3 apply.

Exercise. Show that operators in $\boldsymbol{A} \cup \boldsymbol{B}$ are "deterministic": that is, if f belongs to G, $f = Ef$ (i.e., f is deterministic), and if D is in $\boldsymbol{A} \cup \boldsymbol{B}$, then $Df = EDf$.

Exercise. Let h be a fixed element of G_0, $0 = t_0 < t_1 < \cdots < t_n = T$, $c_1, c_2, \cdots, c_n$ any elements of X with $E(c_i) = 0$, and M a continuous linear transformation of X satisfying $Mh(t_p) = c_p$ for $p = 1, 2, \ldots, n$. Define D on G by

$$[Df](t) = \begin{cases} 0 & -r \le t \le 0 \\ (t - t_{p-1})Mf(t_{p-1}) + [Df](t_{p-1}) & t_{p-1} < t \le t_p \end{cases}$$

Show that D belongs to $\boldsymbol{B}$.

Exercise. As shown in Section 2 of Chapter I, elements of $\boldsymbol{A} \cup \boldsymbol{B}$ are causal: $P_t D = P_t D P_t$ for each t in S and D in $\boldsymbol{A} \cup \boldsymbol{B}$. Find an operator D, of the form given in the preceding exercise, to disprove the following: if f in G_0 is nonanticipating, then Df is nonanticipating.

We recall the operator C in $\boldsymbol{B}$, given by $[Cf](t) = \int_{[0,t]} f(s)\, dk(s)$ for $t > 0$ and

$[Cf](t) = 0$ for $t \leq 0$. Note that $[C^*g](t) = k(t)g(T) - [P_0k](t)g(0) - [Cg](t)$ gives the adjoint of C with respect to the Hilbert space $\{G_H, Q_H\}$, as defined below.

As in previous chapters, G_H contained in G is the space of Hellinger integrable functions for which $\int_{[-r,T]} | df |^2 /dk < \infty$, and the inner product Q_H is given by $Q_H(f, g) = \langle f(-r), g(-r)\rangle + \int_{[-r,T]} \langle df, dg\rangle /dk$. The function K from $S \times S$ into the linear transformations of X defined by $K(s, t) = \min(k(s), k(t))I$, where I is the identity transformation on X, is a reproducing kernel for the Hilbert space $\{G_H, Q_H\}$.

Noise processes. In the first part of this section, the spaces necessary to analyze stochastic hereditary systems $h = Bh + z$ with B in $\boldsymbol{B}$ and z in G_0, were defined. Before solving control problems for such systems, we discuss two noise processes which have proved to be of special importance in applications: namely, the Wiener process W and the Poisson process N. The processes W and N are canonical examples of continuous and point (jump) processes, respectively. Their properties should guide the reader's intuition about the general noise term z belonging to G_z.

For the sake of discussion, suppose that $h = Bh + z$ models a mechanical system. The physics of the system is known and $h = Bh$ describes the ideal relationships. However, a multitude of minor imperfections (wear, dirt, etc.) affect the operation of the system. The noise term z is added to the model to compensate for the system flaws. What properties should z have?

We start by reviewing the Wiener model of Brownian motion. Small particles in water were observed under microscopes to follow wildly erratic paths, as the result of constant molecular bombardment by the liquid. Let z describe the path of a particle, and consider the displacement $\Delta z = z(t + \Delta t) - z(t)$ for some $\Delta t \geq 0$. If the molecular collisions are occuring rapidly and from all angles throughout the liquid, it is reasonable to guess the following: the collisions will tend to cancel each other (i.e., $E(\Delta z) = 0$), Δz does not depend on the location z(t) or on previous values of z (i.e., z has independent increments), and Δz is a normally distributed random variable (the Central Limit Theorem of statistics states that the sum of n independent identically distributed random variables of any type approaches a normally distibuted random variable as $n \to \infty$). Our "guesses" were verified by Einstein in 1905 [7] and a full description of the mathematical model of Brownian motion was given by Wiener in 1920 [26].

A Wiener process (also called Brownian motion) may be defined as follows. A continuous one dimensional process W is a Wiener process on [0, T] if $W(0) = 0$, W has independent increments, and there exists a constant σ such that $W(t) - W(s)$ is normally distributed with mean 0 and variance $\sigma^2(t - s)$ whenever $0 \leq s \leq t \leq T$. A d-dimensional process is a Wiener process if each component is a one dimensional Wiener process. If $\sigma = 1$, W is called standard.

As indicated above, the Wiener process may be a valuable modeling tool if many small disturbances affect a system. Also, W is a simple example of a fractal, whose importance in modeling is vividly demonstrated by Mandelbrot [8].

Exercise. For $r = 0$, show that W belongs to G_z. For $r \neq 0$, define $W(t) = 0$ for $-r \leq t \leq 0$, and show that W belongs to G_z.

Exercise. In this problem, we show that W may be viewed as a limit of a random walk (see Lin and Segel [6] for a more thorough discussion of this viewpoint). Suppose a man starts at position 0 at time 0. At each of the times Δt, $2\Delta t$, $3\Delta t$, $\cdots$, the man flips a coin. If the coin comes up heads, the man takes a step of size Δx to the left; if the coin comes up tails, the man takes a step of size Δx to the right. Thus he moves left with probability 1/2 and moves right with probability 1/2. We are interested in $w(m, N)$ = probability that the man is at position $m\Delta x$ at time $N\Delta t$, and $\underline{w}(x, t) = w(x/\Delta x, t/\Delta t)$ where $x = m\Delta x$ and $t = N\Delta t$. Verify $\underline{w}(x, t + \Delta t) = (1/2)\underline{w}(x - \Delta x, t) + (1/2)\underline{w}(x + \Delta x, t)$, $\underline{w}(0, 0) = 1$, $\underline{w}(x, 0) = 0$ for $x \neq 0$. It can be shown that $w(m, N) = (N!2^N/p!)(N - p)!$ provides the solution to the above difference equation, but we are interested in letting Δx and Δt go to 0. With this in mind, expand the difference equation in a Taylor series about (x, t) to show $v_t(x, t)\Delta t + 0(\Delta t^2) = (1/2)v_{xx}(x, t)\Delta x^2 + 0(\Delta x^3)$, where v is a smooth function agreeing with $\underline{w}$ at each point (m, N). Finally, let $u = (v/2)\Delta x$ and $\Delta x^2/(2\Delta t) = D$, and show that $u_t = Du_{xx}$ at each point (m, N). Explain why $\int_{[a,b]} u(x, t)\, dx$ gives the probability of finding the man in the interval [a, b] at time t. Show that $u(x, t) = (4\pi Dt)^{-1/2} \exp(-x^2/4Dt)$ satisfies $u_t = Du_{xx}$, $\int_{[-\infty,\infty]} u(x, t)\, dx = 1$ and $\lim_{t\downarrow 0} u(x, t) = 0$, $x \neq 0$. This indicates that u is the probability density function of a continuous random variable. It is, in fact, the density function of a normal random variable with mean 0 and variance 2Dt. Since the random walk clearly has independent increments, our limit-

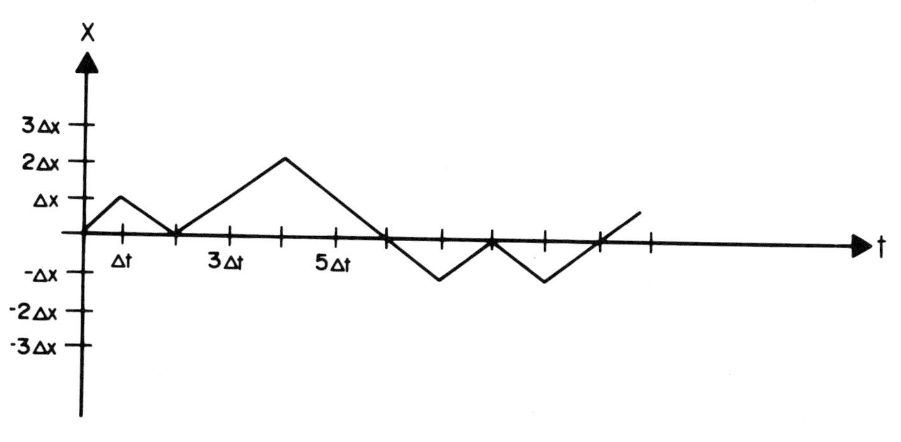

Figure 4.2. A sample path of a random walk.

ing process must be a (nonstandard, unless $D = 1/2$) Wiener process. It is a fact that W (viewed as $W(\,, \omega)$ for fixed ω in Ω) is non differentiable almost everywhere, which is difficult to prove but which is made believable by the above construction of W. In spite of the non differentiability of W, a formal derivative of W, called "white noise", is often used in applications. Mathematically, white noise is characterized by a constant spectral function.

Returning to the mechanical system $h = Bh + z$, suppose that the Wiener process does not completely model the system noise. For instance, the magnitude of the disturbance might depend on the state of the system; e.g., a particular segment of a cycle is executed poorly. This case of state dependent noise is discussed in Section 4 of this chapter.

Another situation in which W might inadequately model the disturbance is if the system is subjected to recurrent shocks of significant size. The Poisson process, defined below, is commonly used to model events occurring at discrete random times.

Suppose that $N(t)$ is the number of events (in our case, shocks) observed from time 0 until time t, with $N(0) = 0$. The events occur randomly under the following rules:

1. At most one event occurs at any time t.
2. The probability of more than one event occurring between times t and $t + \Delta t$ is $0(\Delta t)$ as $\Delta t \to 0$.
3. The probability of one event occurring between t and $t + \Delta t$ is $\lambda\Delta t + 0(\Delta t)$ as $\Delta t \to 0$ for some $\lambda > 0$.

The (right continuous) stochastic process N(t) is called a stationary Poisson process. The constant λ is called the rate of the process since $E(N(t)) = \lambda t$; furthermore, $1/\lambda$ is the expected length of time between events. The properties (1) - (3) imply that the Poisson process models events occuring at isolated times with the "memoryless" property that the past $\{N(s) : 0 \le s \le t\}$ does not affect the future $\{N(s) : s > t\}$.

Exercise. Show that $N(t) - \lambda t$ belongs to G_z for r = 0.

Exercise. For events obeying (1) - (3), let $P_n(t)$ be the probability that n events occur in [0, t]. Show that $P_0(t + h) = P_0(t)(1 - \lambda h) + o(h)$. Letting $h \to 0$, show $dP_0(t)/dt = -\lambda P_0(t)$, and hence $P_0(t) = e^{-\lambda t}$. Similarly, $P_n(t + h) = P_n(t)(1 - \lambda h) + P_{n-1}(t)\lambda h + o(h)$. Letting $h \to 0$, obtain $dP_n(t)/dt = -\lambda P_n(t) + \lambda P_{n-1}(t)$. Using induction, conclude $P_n(t) = (\lambda t)^n e^{-\lambda t}/n!$

3. Control of Systems with State Independent Noise

In this section, we analyze stochastic hereditary systems of the form $h = Bh + u + z$, with B in ***B*** and z in G_0. The random input z represents state independent noise, and u in G_H is the control function to be determined. The system dynamics is equivalently written $h = A(u + z)$, $A = (I - B)^{-1}$.

Preliminary results. Our goal is to find the feedback control $u = Dh$, for some D in ***B***, which minimizes a given cost functional. The general procedure is to formulate the control problem as a constrained optimization problem in terms of the Hellinger RKH space, apply the Lagrange Multiplier Theorem, rewrite the functionals in the Lagrange necessary condition in terms of Q_H, apply a version of the DuBois-Reymond Lemma, use the constraints to obtain a necessary condition for D (and hence u), and show that the necessary condition is also sufficient. To simplify later arguments, we pause now to prove results which are basic to the third and fourth of these steps.

For the problems to be considered below, the Lagrange Multiplier Theorem (discussed in Chapter II) gives a necessary condition containing terms of the form $\lambda_1(f)$, where λ_1 is an unspecified continuous linear functional on G and f is the variation of the constraint. Our immediate concern is to write $\lambda_1(f)$ in terms of the Hellinger inner product Q_H. We note that, for f in G_H, Riesz-Fischer (Theorem 1.5) guarantees there is a λ in G_H such that $\lambda_1(f) = Q_H(f, \lambda)$.

However, since λ_1 is defined on all of G, the form of Riesz-Fischer that applies is $\lambda_1(f) = \int_S <f, d\lambda_2>$, where λ_2 is a function of bounded variation which may be chosen to have the value $\lambda_2(T) = 0$. Performing integration by parts, we get

$$\lambda_1(f) = \int_{-r}^{T} <f, d\lambda_2> = <f(T), \lambda_2(T)> - <f(-r), \lambda_2(-r)> - \int_{-r}^{T} <df, \lambda_2>$$

$$= - <f(-r), \lambda_2(-r)> - \int_{-r}^{T} <df, \lambda_2>$$

To relate the above expression to Q_H, we let $f = K(\cdot, t)x$ for some t in S and x in X, and discover that

$$\lambda_1(K(\cdot, t)x) = -<K(-r, t)x, \lambda_2(-r)> - \int_{-r}^{T} <dK(\cdot, t)x, \lambda_2>$$

$$= - <x, \lambda_2(-r)> - <x, \int_{-r}^{t} \lambda_2 \, dk>$$

$$= Q_H(K(\cdot, t)x, -\lambda_2(-r) - C\lambda_2)$$

Since K is the reproducing kernel for $\{G_H, Q_H\}$, it can be shown (see Exercise below) that $\lambda_1(f) = Q_H(f, -\lambda_2(-r) - C\lambda_2)$ for any f in G_H. We have proved the following.

<u>Lemma</u> 4.1. If λ_1 is a continuous linear functional on G, then there exist λ_2 of bounded variation and $\lambda \in G_H$ such that;

1. $\lambda_1(f) = \int_S <f, d\lambda_2>$ for any f in G.
2. $\lambda_1(f) = Q_H(f, \lambda)$ for any f in G_H, with $\lambda = -\lambda_2(-r) - C\lambda_2$.

<u>Exercise</u>. Show that part (1) of Lemma 4.1 follows from Riesz-Fischer.

Exercise. Show that if $\int_S \langle f, d\lambda_2 \rangle = Q_H \langle f, -\lambda_2(-r) - C\lambda_2 \rangle$ for any f of the form $f = K(\bullet, t)x$ with t in S and x in X, then the formula holds for all f in G_H.
HINT: $\{\Sigma_{p=1,n} K(\bullet, t_p)x_p : t_p \in S, x_p \in X\}$ is dense in G_H.

We next consider the following problem. If $Q_H(h, f) = 0$ for all h in G_H, or $Q_H(Dh, f) = 0$ for all D in $\boldsymbol{B}$ (and fixed h in G), does it follow that $f = 0$? The answer is obviously yes in the first case, since we may choose h = f and $Q_H(f, f) = 0$ implies f = 0. For the second case, we need a subtler version of the DuBois-Reymond Lemma.

Lemma 4.2. If A belongs to $\boldsymbol{A}$, z belongs to G_0, h = Az and f is in G_0 and G_H, then $Q_H(Dh, f) = 0$ for all D in $\boldsymbol{B}$ implies f = 0.

Proof of Lemma 4.2. It suffices to show that $\{f = Dh : D \in \boldsymbol{B}\}$ is dense in the intersection of G_H and G_0. First, we note that h belongs to G_0, so that Dh belongs to G_H and G_0, for any D in $\boldsymbol{B}$. Next, let $0 = t_0 < t_1 < \ldots < t_N = T$ and suppose $\{c_p\}_{p=1,N}$ is any sequence of elements in $L_{2,d}(\Omega)$ with $Ec_p = 0$ for each p. By the Hahn-Banach Theorem, there exists a continuous linear transformation M of $L_{2,d}(\Omega)$ such that $M[h(t_p)] = c_p$ for each p. Fix M and define a linear transformation D_N on G by

$$[D_N f](t) = \begin{cases} 0 & t \leq 0 \\ (k(t) - k(t_{p-1}))M[h(t_{p-1})] + [D_N f](t_{p-1}) & t_{p-1} < t \leq t_p \end{cases}$$

Note that $[D_N h](0) = [D_N h](t_1) = 0$, $[D_N h](t_2) = (k(t_2) - k(t_1))c_1$, etc. Thus $D_N h$ is a K-polygon (piecewise linear if k is linear). Furthermore it was verified in the Exercises above that $ED_N h = 0$ and D_N is in B, so that $D_N h$ belongs to G_H and G_0. A property of any RKH space is that K-polygons are dense in space. Since N, $\{t_p\}_{p=1,N}$, and $\{c_p\}_{p=1,N}$ are arbritrary, $\{f = DH: D \text{ in } \boldsymbol{B}\}$ contains all K-polygons in G_H and G_0 and hence is dense in the intersection of G_H and G_0. □

Control of state space systems. Before considering control of the general stochastic hereditary system h = Bh + u + z, we look at the state space case where $B = \alpha C$ for some constant matrix α. Then, if u and z are differentiable, system dynamics are given by the first order differential equation $h' = \alpha h + u' + z'$. State space systems are analyzed separately for two reasons. They are of special importance in applications, and the simple form of B enables us to minimize a cost functional of a particularly nice form.

Along with taking $B = \alpha C$, in the state space case we fix $r = 0$ and $dk(t) = dt$. The objective of the control problem is to choose a feedback control $u = Dh$ for some D in $\boldsymbol{B}$ to minimize

$$J(D, h) = (1/2)E[\int_0^T (c_1 \mid h(t) \mid^2 + c_2 \mid du/dt \mid^2)\, dt + \mid h(T) \mid^2]$$

The cost functional J is quadratic in h, du/dt, and the final state h(T). The penalty assigned to du/dt may seem natural when thinking of the system dynamics as the differential equation given above. As noted below, this is a standard cost functional in applications.

For a precise formulation of the problem, the first step is to rewrite the cost functional in terms of Q_H. It is clear that

$$J(D, h) = (1/2)\{c_1 N_H^2(Ch) + c_2 N_H^2(Dh) + Q_H(h, kh(T))\}$$

with the control written explicitly as $u = Dh$. Since $Ez = 0$ and we require $u = Dh$, it follows that $h = (I - \alpha C - D)^{-1}z$ and h belongs to G_0. Our constrained optimization problem is to find (D, h) in $\boldsymbol{B} \times G_0$ which minimizes J subject to $h = \alpha Ch + Dh + z$.

The Lagrange Multiplier Theorem applies, and gives the following necessary condition for a minimizing pair (D, h): there exists a continuous linear functional λ_1 on G_0 such that for every $(D', h') \in \boldsymbol{B} \times G_0$,

$$\delta J(D, h; D', h') + \lambda_1 (\delta M(D, h; D', h')) = 0$$

where M is the constraint $M(D, h) = h - \alpha Ch - Dh - z = 0$ and the variations δJ and δM of J and M, respectively, are given by

$$\delta J(D, h; D', h') = c_1 Q_H(Ch', Ch) + c_2 Q_H(D'h+Dh', Dh) + \langle h'(T), h(T)\rangle$$
$$\delta M(D, h; D', h') = h' - \alpha Ch' - D'h - Dh'$$

Using Lemma 4.1 to rewrite $\lambda_1(\delta M)$, the necessary condition becomes: there exists a λ in G_H such that for every (D', h') in $\boldsymbol{B} \times G_0 \cap G_H$,

$$c_1 Q_H(Ch', Ch) + c_2 Q_H(D'h + Dh', Dh) + Q_H(h', kh(T)) + Q_H(h' - \alpha Ch' - D'h - Dh', \lambda) = 0 \tag{4.1}$$

Special cases of the basic relation (4.1) enable us to derive a more precise necessary condition, as well as providing a sufficiency argument. If $h' = 0$, (4.1) reduces to $Q_H(D'h, c_2Dh-\lambda) = 0$ for each D' in $\boldsymbol{B}$. Then, Lemma 4.2 implies that

$$\lambda = c_2 Dh \tag{4.2}$$

since $[Dh](0) = \lambda(0) = 0$. On the other hand setting $D' = 0$ in (4.1) and using the Dubois-Reymond Lemma yields

$$(I - P_0)\{c_2 D^*Dh + c_1 C^*Ch + kh(T) + (I - \alpha^*(C^* - D^*)\lambda\} = 0 \tag{4.3}$$

Substituting (4.2) into (4.3), we get the necessary condition

$$(I - P_0)\{c_1 C^*Ch + kh(T) + (I - C^*\alpha^*)c_2 Dh\} = 0 \tag{4.4}$$

We will show below that (4.4) is also a sufficient condition for a minimizing pair (D, h). First, however, we put (4.4) into a more practical form.

<u>Exercise</u>. Substituting for C^*, show that (4.4) may be written $\lambda + \alpha^*C\lambda = c_1C^2h + (k - 1)\gamma$, with $\gamma = -c_1[Ch](T) - h(T) + \alpha^*\lambda(T)$.

Using the above Exercise, we have

$$\lambda = (I + \alpha^*C)^{-1}(c_1C^2h + (k - 1)\gamma) \tag{4.5}$$

which must be reconciled with (4.2). Intuitively, since γ is constant and k is deterministic, $\lambda = c_2Dh$ for some D in $\boldsymbol{B}$ must imply $\gamma = 0$. This is verified below.

From (4.5), with $\lambda = c_2 Dh$, we see $(I + \alpha^*C)^{-1}(k - 1)\gamma = c_2Dh - (I + \alpha^*C)^{-1} c_1C^2h = D_1h = B_1z$, where $D_1 = c_2D - (I + \gamma^*C)^{-1}c_1C^2$ and $B_1 = D_1 (I - \alpha C\text{-}D)^{-1}$. In the last equality, we have applied the constraint $h = (I - \alpha C - D)^{-1}z$. Then, since we have fixed $k(t) = 1 + t$, $\gamma = [(I + \alpha^*C)B_1z](t)/t$ for t in (0, T]. Taking a limit of the above equation as $t \to 0$, and observing that $(I + \alpha^*C)^{-1}(k - 1)\gamma$, and hence B_1z, is differentiable, L'Hopital's Rule gives $\gamma = \lim_{t\downarrow 0}[(I + \alpha^*C)B_1z](t)/t = \lim_{t\downarrow 0}\{B_1z]'(t) + \alpha^*[B_1z](t)\} = [B_1z]'(0) = \lim_{t\to 0}[B_1z)(t)/t$. If c is the constant corresponding to B_1 in ***B***, then

$$| [B_1z](t)/t | \le (c/t) \int_0^t N_\tau(z)d\tau \le cN_t(z) \to 0 \qquad \text{as } t \to 0$$

Thus, $\gamma = 0$ and the necessary condition is

$$u = Dh = \lambda/c_2 = (c_1/c_2)(I + \alpha^*C)^{-1}C^2h \tag{4.6}$$

It is an important feature of our method that we may show that (4.4) is also a sufficient condition for the solution of the control problem. The basic idea is to work backwards, and it is easy to see that (4.4) implies that (4.1) holds for every (D', h') in $B \times G_0 \cap G_H$ and $\lambda = c_2Dh$. We adopt the following notation: (D, h) satisfies (4.4) with $h = \alpha Ch + Dh + z$, and (D', h') in $B \times G_0$ satisfies $h' = \alpha Ch' + D'h' + z$.

Exercise. Show that $h - h'$ belongs to $G_0 \cap G_H$. Applying (4.1) to $(D', h - h')$, show that

$$\begin{aligned} c_1N_H^2(Ch) + c_2N_H^2(Dh) &= c_1Q_H(Ch' - Ch) + (1/2)c_2N_H^2(D'h' - Dh) + (1/2)\,|\,h'(T) - h(T)\,|^2 \\ &= J(D', h') + J(D, h) - c_1Q_H(Ch', Ch) - c_2Q_H(D'h', Dh) - \langle h'(T), h(T)\rangle \\ &= J(D', h) + J(D, h) - c_1N_H^2(Ch) - c_2N_H^2(Dh) - Q_H(h, kh(T)) \\ &= J(D', h') - J(D, h) \end{aligned}$$

Thus, $J(D,h) \le J(D', h')$ for any pair (D', h') satisfying the constraints, with equality only if $h' = h$ and $D' = D$. The above work is summarized in the following result.

Theorem 4.3. A necessary and sufficient condition for (D, h) in $\boldsymbol{B} \times G_0$ to minimize $J(D,h) = (1/2)c_1N_H(Ch)^2 + (1/2)c_2N_H(Dh)^2 + (1/2)\,|\,h(T)\,|^2$ subject to $h = \alpha Ch + Dh + z$, for a fixed constant matrix α, is that $(I - P_0)\{c_1C^*Ch + c_2(I - \alpha^*C^*)Dh + kh(T)\} = 0$, where C^* is the adjoint of C in $\{G_H, Q_H\}$ and α^* is the matrix transpose of α. Furthermore, if the above condition holds, then the optimal control is given by $u = Dh = (c_1/c_2)(I + \alpha^*C)^{-1}C^2h$.

We emphasize a few of the characteristics of the optimal control determined above. First, u does not depend directly on the statistics of z, and the feedback operator D may be computed analytically without solving a Riccati differential equation. In fact, $(I + \alpha^*C)^{-1}$ may be computed using Laplace transforms, as shown in the example to follow. Also, the magnitude of u is proportional to c_1/c_2, the ratio of the weights assigned to the output and control costs, respectively. Finally, although T appears in the necessary and sufficient condition (4.4), u is independent of the terminal time T.

The results of Theorem 4.3 are now applied to a pendulum model adapted from Russell [24]. The physical situation to be referred to is that of a pendulum which is constrained to move in a one dimensional arc. The pendulum is thought of as a surveyor's instrument subjected to a random force w, which may represent wind or other disturbances. The connection of the pendulum to its housing causes friction, so that a simple linear model of the displacement angle $\theta(t)$ is

$$\theta'' + \theta' + \theta = w \qquad \theta(0) = a \qquad \theta'(0) = b \tag{4.7}$$

where a and b are mean-zero random variables. A control function is to be chosen as a function of θ and θ', and is allowed to directly alter θ and θ'.

To apply most state space methods, it is necessary that the system noise be gaussian. This requirement may play a large role in determining the state vector. For example, Russell uses the state vector $X = (\theta, \theta', w)^*$, because wind is not well modeled by white noise (under a white noise model, $w(t + \Delta t)$ is independent of $w(t)$ for any $\Delta t > 0$). If w is driven by white noise (that is, $w' + w = v$ = white noise), then the system is indeed linear and gaussian:

$$X' = \begin{bmatrix} 0 & 1 & 0 \\ -1 & -1 & 1 \\ 0 & 0 & -1 \end{bmatrix} X + \begin{bmatrix} 0 \\ 0 \\ 1 \end{bmatrix} v \qquad (4.8)$$

If w is not driven by white noise, further components (e.g., w′) might be added to the state vector. However, since the control only affects θ and θ', components other than θ and θ' are in a practical sense superfluous.

Removing the gaussian restriction on the system noise allows us to use the "natural" state vector $h = (\theta, \theta')^*$. The resulting model is:

$$h = \alpha Ch + z \qquad \alpha = \begin{bmatrix} 0 & 1 \\ -1 & -1 \end{bmatrix} \qquad z(t) = (0 \int_0^t w(s)\,ds)^* \qquad (4.9)$$

Exercise. Show that (4.7) and (4.9) are equivalent if a = b = 0. If w has mean zero, show that z belongs to G_0. Under what conditions on w is z in G_z? Hint: See the discussion preceding (4.8).

Exercise. Show that (4.7) may also be written $\theta + C\theta + C^2\theta = C^2w + f$ where f depends on $\theta(0)$ and $\theta'(0)$. If $B = -C - C^2$ and $z = C^2w + f$, this model is of the form $\theta = B\theta + z$. Show that B belongs to ***B***. What conditions on w and f imply z belongs to G_0 or G_z?

The control function (4.6) and the resulting output $h = (I - \alpha^*C)^{-1}f$ may be computed using Laplace transforms. If $g = (I + \alpha^*C)^{-1} f$, $G = L(g)$, and $F = L(f)$ then $(I + \alpha^*/s)\,G = F$ and hence

$$g(t) = f(t) + \int_0^t m(t-u)f(u)du$$

$$m(x) = (1/3)\exp(x/2)\sin(\sqrt{3}\,x/2)\begin{bmatrix} -2 & 1 \\ -1 & -1 \end{bmatrix} + \cos(\sqrt{3}\,x/2)\begin{bmatrix} 0 & 1 \\ -1 & 1 \end{bmatrix}$$

A similar computation yields

$$\theta(t) = \int_0^t n(t-u)w(u)\,ds$$

$$n(x) = \cosh(\lambda x/2)(\cos\beta x - (1/2)\sin\beta x) + \sinh(\gamma x/2)\{(5\beta\gamma/2)\sin\beta x - (1/\gamma)\cos\beta x\}$$

$$\gamma = \sqrt{1+2\sqrt{5}} \qquad ß = \sqrt{2\sqrt{5}-1/2}$$

from which it is relatively easy to compute $E(\theta^2)$.

Exercise. Verify the above computations.

Other methods for computing $E(\theta^2)$ will be given later in this section. One method is based on the RKH space theory corresponding to covariance kernels. A state space method of Russell is also given in the Summary closing the Chapter.

Control of general hereditary systems. The state space system discussed above was characterized by the hereditary dynamics $h = Bh + u + z$, with B in ***B*** specified to be of the form $B = \alpha C$. The simple form of B gave us several analytical advantages, the most important of these being the explicit, and convenient representation of the adjoint B^*. Having B^* in hand enabled us to derive the control function (4.6) from the necessary condition (4.4).

The general stochastic hereditary system $h = Bh + u + z$, with B in ***B*** unspecified, $u = Dh$ for some D in ***B*** to be chosen, and z in G_0, is examined below. There are no difficulties determining a necessary and sufficient condition analogous to (4.4) for the cost functional J(D, h) defined above. The complications lie in the derivation of a suitable control law from the necessary condition. To accomplish this for general B in ***B***, we modify the cost functional to be minimized. The resulting necessary condition is then manipulated to extract the control function.

We now state the control problem to be solved below. As in the state space case, the control is explicitly written as $u = Dh$. The operator B in ***B*** and noise z in G_0 are fixed but arbitrary, and $A = (I - B)^{-1}$. We wish to find the pair (D, h) in $\boldsymbol{B} \times G_0$ that minimizes

$$J(D, h) = (c_1/2)N_H^2(Au) + (c_2/2)N_H^2(Ch) + (1/2)\,|\,h(T)\,|^2 \tag{4.10}$$

subject to $h = Bh + Dh + z$.

The modified cost functional, aside from being convenient analytically, may arise naturally in some applications. The change from the standard cost functional introduced in the state space problem is the substitution of Au for u in the first term of (4.10). Noting that $h = Au + Az$, we see that Au is the controlled part of h. That is, Au is the difference between the controlled and uncontrolled systems. Instead of penalizing changes in the control u, then, we are now penalizing changes in the system response to the control. This may be appropriate if the implementation of the control is inexpensive and large changes in state are to be avoided; e.g., steering an airplane with passenger comfort of paramount importance.

Before solving the control problem we briefly examine one of the important examples of a hereditary system. The operator B in $\boldsymbol{B}$ may be composed in part of operators of the form $[B_r](t) = \int_{[0,t]} f(t, s)h(s - r)\, ds$, $t > 0$, for some delay $r > 0$. The assumption that z belongs to G_0 gives the equation $h = B_r h + z$ a different look than usual. For delay equations, the solution h is prescribed on an initial interval of length r. The effect of requiring that z belongs to G_0 is to shift the initial interval to $[0, r]$.

<u>Exercise</u>. Consider the equation $h = B_r h + z$ where z in G_0 and B_r is the delay operator given above. On which intervals are the following true: $h = 0$? $h = z$?

We return to the constrained optimization problem given above. As in the state space case, our starting point is the Lagrange Multiplier Theorem. The Lagrange necessary condition for a minimizing pair (D, h) in $\boldsymbol{B} \times G_0$ is that there exists a continuous linear functional λ_1 on G_0 such that for every (D′, h′) in $\boldsymbol{B} \times G_0$,

$$\delta J(D, h; D', h') + \lambda_1(\delta M(D, h; D', h')) = 0$$

where J is given by (4.10) and the constraint M is $M(D, h) = h - ADh - Az = 0$. We compute the variations

$$\delta J(D, h; D', h') = c_1 Q_H(ADh' + AD'h, ADh) + c_2 Q_H(Ch', Ch) + \langle h'(T), h(T)\rangle$$
$$\delta M(D, h; D', h') = h' - ADh' - AD'h$$

Lemma 4.1 now applies, since AD′ belongs to $\boldsymbol{B}$ and any D′′ in $\boldsymbol{B}$ may be written as $D'' = AD'$ with $D' = A^{-1}D''$ in $\boldsymbol{B}$. Then, if h′ belongs to G_H, the above necessary condition becomes: there exists λ in G_H such that

$$c_1 Q_H(AD'h + ADh', ADh) + c_2 Q_H(Ch', Ch) + Q_H(h', kh(T)) + Q_H(h' - AD'h - ADh', \lambda) = 0 \tag{4.11}$$

for each (D', h') in $B \times G_0 \cap G_H$. We consider the special cases $D' = 0$ and $h' = 0$ in (4.11), and use Lemma 4.2 to deduce the following necessary conditions for λ, D, and h.

$$(D'=0) \quad (I - P_0)\{c_1(AD)^*ADh + c_2C^*Ch + kh(T) + (I - AD)^*\lambda\} = 0$$
$$(h'=0) \quad c_1ADh - \lambda = (I - P_0)\{c_1ADh - \lambda\} = 0$$

We substitute $\lambda = c_1ADh$ into the top equation, and get

$$(I - P_0)\{c_2C^*Ch + kh(T) + c_1ADh\} = 0 \tag{4.12}$$

An argument similar to that employed in the state space case may be used to show that (4.12) is also a sufficient condition. We note the similarity of (4.12) and (4.4), with the important difference that (4.12) does not contain a B* or A* term. We may thus derive a simple form for the control function.

Exercise. Show that

$$c_1N_H^2(Dh) + c_2N_H^2(Ch) - c_1Q_H(Dh, D'h') - c_2Q_H(Ch, Ch') + Q_H(h-h', kh(T)) = 0$$

for (D, h) satisfying (4.12) and any (D', h') in $\boldsymbol{B} \times G_0$ with $h' = Bh' + D'h' + z$. Then show that

$$J(D', h') - J(D, h) = (c_1/2)N_H^2(D'h' - Dh) + (c_2/2)N_H^2(Ch' - Ch) + (1/2)\,|\,h'(T) - h(T)\,|^2$$

and hence (4.12) is a sufficient condition.

To derive the control law, we expand C^*Ch in (4.12) and, defining $-\gamma = c_2[Ch](T) + h(T)$, obtain $c_1ADh = c_2C^2h + \gamma(I - P_0)k$. We conclude that, if $\gamma = 0$,

$D = (c_2/c_1)A^{-1}C^2 = (c_2/c_1)(I - B)C^2$. To show $\gamma = 0$, we start with the constraint $K = h - ADh - Az = 0$. Then

$$h = ADh + Az = (c_2/c_1)C^2h + (\gamma/c_1)(I - P_0)k + Az$$
$$= \{(c_2/c_1)C^2(I - AD)^{-1} + A\}z + (\gamma/c_1)(I - P_0)k$$

Solving for γ, we get

$$\gamma = c_1\{(I - AD)^{-1} - (c_2/c_1)C^2(I - AD)^{-1} - A\}z/(I - P_0)k = B_1z/(I - P_0)k$$

The algebraic properties of ***A*** and ***B*** imply B_1 belongs to ***B***. Then, if k is right differentiable at 0 (we have previously assumed right continuity) with derivative $s \neq 0$, $\gamma = \lim_{t\downarrow 0}[B_1z](t)\{k(t) - k(0)\} = [B_1z]'(0)/s = 0$. We have proved the following.

Theorem 4.4. A necessary and sufficient condition for (D, h) in $B \times G_0$ to minimize $J(D,h) = (1/2)c_1N_H(ADh)^2 + (1/2)c_2N_H(Ch)^2 + (1/2)\,|\,h(T)\,|^2$ subject to $h = A(Dh + z)$ is that $(I - P_0)\{c_2C^*Ch + kh(T) + c_1ADh\} = 0$, where C^* is the adjoint of C in $\{G_H, Q_H\}$. Furthermore, if k is right differentiable at 0 with derivative $s \neq 0$ and the above condition holds, then $u = Dh = (c_2/c_1)(I - B)C^2h$.

To illustrate the above result, we consider the pendulum example given at the end of the sub-section on state space control. By reworking this example, we also allow for comparisions between the state space and general hereditary control problems in the absence of delays.

From a modeling viewpoint, the general theory has the advantage that we are not compelled to use (4.9), although we may. An alternative is to integrate (4.7) twice and get the (uncontrolled) equation $h = Bh + z + f$, where $B = -C - C^2$, $z = C^2w$, and f depends on a and b. In this case, the state is $h = \theta$, so that the control will only be dependent on the angle θ, and not explicitly affect the velocity.

The computations are a little more straightforward here than in the state space case. If $c_1 = c_2$, the control is given by $u = (C^2 + C^3 + C^4)h$. The Laplace transform technique discussed previously yields

$$h(t) = [(I - B - (I - B)C^2)^{-1}z](t0$$
$$= z(t) - (g*z)(t)$$
$$g\,(s) = e^{-s}/3 + 2e^{-s/2}\cos(\sqrt{3}\,s/2)/3$$

where $g*z$ denotes the normal convolution of g and z. As in the state space case, we are able to compute the output h explicitly, and may thus compute the covariance $E(h^2)$ directly.

Exercise. Verify the above computations.

Exercise. Compute the optimal control $u = (I - B)C^2h$ using the two dimensional model (4.9), and compare the resulting output h with the value of θ obtained from implementing the state space control.

Exercise. Suppose that B belongs to $\boldsymbol{B}$ and $B^*f = B'f + k\gamma$, where B' belongs to $\boldsymbol{B}$ and γ is a constant (possibly depending of f). Find $(D, h) \in \boldsymbol{B} \times G_0$ to minimize $J(D, h) = (1/2)c_1N_H(Ch)^2 + (1/2)c_2N_H(Dh)^2 + (1/2)\,|\,h(T)\,|^2$ subject to $h = Bh + Dh + z$. Does $B = -C - C^2$ satisfy the above condition?

A time-series approach to covariance computations. To this point, we have obtained results similar to the deterministic results of Chapter II, using predominantly the same theoretical methods. While believing this to be an advantage of our approach, we also recognize the need to approach some problems from a uniquely stochastic viewpoint. Below, we solve a standard systems theory problem by borrowing some ideas from time series analysis.

It is common in the systems theory literature to make a "mean zero" assumption: the output h(t) has expected value 0 at each time t. For the dynamics $h = Bh + u + z$ with B in $\boldsymbol{B}$, the assumptions z in G_0 and $u = Dh$ with D in $\boldsymbol{B}$, imply $Eh = 0$. Since the mean (first moment) of h is 0, the main measure of the size of the output is the covariance (second moment) $R(s, t) = Eh(s)h(t)^*$, where * denotes matrix transpose. Our goal is to write the (output) covariance kernel of $h = Az$ with A in $\mathcal{A}$, z in G_z, in terms of the operator A and the covariance of the noise process.

In the discussion to follow, we assume that w in G is a d-dimensional stochastic process with covariance kernel R_w. It is easy to see that R_w is positive, in the sense that $\Sigma_{i,j=1,N} v_i^* R_w(t_i, t_j) v_j \geq 0$ whenever $\{t_i\}_{i=1,N}$ is in S and $\{v_i\}_{i=1,N}$ is in R^d. This follows from $E\{(\Sigma_{i=1,N} v_i^* w(t_i))^2\} \geq 0$. A cornerstone of the theory of reproducing kernels is the equivalence of the following statements [2]: R_w is positive, and R_w is the reproducing kernel for a unique RKH space. Thus, every covariance kernel is the reproducing kernel of a RKH space, which we denote $\{H_w, <\cdot, \cdot>_w\}$.

The reproducing property of R_w implies that

$$<R_w(\cdot, s)e_i, R_w(\cdot, t)e_j>_w = <e_i, R_w(s, t)e_j> = Ew_i(s)w_j(t) \tag{4.13}$$

where e_j is the d-vector with j^{th} component δ_{ij}. That is, the statistics of w may be obtained from an appropriate inner product on H_w.

The relationship between w and H_w may be further developed. Let LS(w) denote the linear span of w; then $LS(w) = \{w = \Sigma_{i=1,N} v_i^* w(t_i)\colon v_i \in R^d, t_i \in S, N$ an integer$\}$. Clearly, LS(w) is contained in $L_2(\Omega)$. Let $L_2(w)$ denote the closure of LS(w) with respect to the $L_2(\Omega)$ norm. Then, $\{L_2(w), <\cdot, \cdot>\}$ is a Hilbert space, with inner product $<x, y>_E = Ex^*y$. The following is a restatement of (4.13).

Proposition. The RKH space $\{H_w, <\cdot, \cdot>_w\}$ represents w in the sense that there is a one to one, inner product preserving map between $\{H_w, <\cdot, \cdot>_w\}$ and $\{L_2(w), <\cdot, \cdot>_E\}$ with $R_w(\cdot, t)e_i$ congruent to $w(t)^*e_i$.

To illustrate the above idea, we find the RKH space representation of z in G_z with S = [0,T] and dimension d = 1. Using the independent increments property of G_z, it can be shown that $R_z(s,t) = Ez(s)z(t) = Ez^2(s)$ if $s \leq t$. Defining $k(s) = Ez^2(s)$, we have the familiar formula $R_z(s, t) = k(\min(s, t))$. If k is right continuous and nondecreasing, and $k(0) = 1$, then R_z is the reproducing kernel for a space of Hellinger integrable functions, and $\{G_H, Q_H\}$ is the RKH space representation of z. However, if z is a Wiener process of S, $k(0) = 0$, and $\{G_H, Q_H\}$ does not represent z.

Exercise. Show that f belongs to H_w if and only if $f = Ew(\cdot)c_f$ for some c_f in $L_2(w)$, and $<f, g>_w = Ec_fc_g$. Hint: Show that if $\{t_p\}_{p=1,\infty} \subset S$, $\{v_p\}_{p=1,\infty} \subset R^d$, $f_n = \Sigma_{p=1,n} Ew(\cdot)w(t_p)^*v_p$, a and $r_n = \Sigma_{p=1,n} w(t_p)^*v_p$, then $\{f_n\}_{n=1,\infty}$ is a Cauchy sequence in H_w if and only if $\{r_n\}_{n=1,\infty}$ is a Cauchy sequence in $L_2(w)$.

Exercise. If z belongs to G_z and $s \leq t$, show that $z(s)z(t)^* = Ez(s)z(s)^*$.

Exercise. In this exercise, we examine the validity of the above assumptions on $k(t) = Ez^2(t)$ that imply $H_z = G_H$. Show that if $s \leq t$, then $k(t) - k(s) = E\{(z(t) - z(s))^2\}$. From this, conclude that k is right continuous and nondecreasing. Compute k if $z(t) = W(t)$ for a Wiener process W; $z(t) = N(t) - \lambda t$ for Poisson process N with rate λ. Show $k(0) = 0$ for any z in G_z.

Our initial attempt to identify $\{G_H, Q_H\}$ as the RKH space representation of z in G_z failed because $k(0) = 0 \neq 1$. The condition $k(0) = 1$ is necessary to guarantee the reproducing property of $k(\min(s,t))$ in $\{G_H, Q_H\}$. One way of forcing $k(0) = 1$ is to define $z(0) = 1$. Aside from contradicting standard definitions of noise processes, this alteration would unduly complicate our previous derivations. It is better to "redefine" G_H and Q_H than z.

We define G_{H0} to be $G_H \cap G_0$. For f, g in G_{H0}, the inner product becomes $Q_{H0}(f, g) = \int_{[0,T]} df dg/ dk$. The kernel $K(s, t) = k(\min)(s, t))$ has the reproducing property and $K(\cdot, t)$ is in G_{H0} if $k(0) = 0$. Thus, $\{G_{H0}, Q_{H0}\}$ is an RKH space and, in fact, is the RKH space representation of z in G_z.

There is one more complication to overcome before completing the realization of $\{H_z, <\cdot, \cdot>_z\}$. The above remarks are valid for dimension $d = 1$, but do not make sense for $d \geq 2$. Fortunately, the higher dimensional cases may be handled in essentially the same way as the scalar case.

Following the outline of the above work, we take z in G_z and note that $R_z(s, t) = k(\min(s, t))$, where $k(t) = Ez(t)z(t)^*$ is now a $d \times d$ matrix. We want k to generate an RKH space of Hellinger integrable functions, but have not yet defined Hellinger integration with respect to a matrix valued function. As shown by Mac Nerney [17], the desired form of G_H is a natural generalization of the "scalar" G_H.

We will define Hellinger integration with respect to M(t), a function from S into the $d \times d$ matrices. We assume that M(t) is symmetric for each $t \in S$, $M(-r) = 0$, M is right-continuous on S, and M is nondecreasing in the sense that if $s \leq t$, $x^*(M(t) - M(s))x \geq 0$ for each x in R^d. A function $f: S \rightarrow R^d$ is Hellinger integrable with respect to M if there exists a constant b such that

$$\sum_{p=1}^{n} \{f(s_p) - f(s_{p-1})\}^*\{M(s_p) - M(s_{p-1})\}^{-1}\{f(s_p) - f(s_{p-1})\} \leq b$$

for each nondecreasing sequence $\{s_p\}_{p=1,n}$ in S. The smallest such b is denoted $\int_S$ <df, dM^{-1}df>. We define G_{H0} to be the space of all Hellinger integrable functions with $P_0 f = 0$. As in the scalar case, $\{G_{H0}, Q_{H0}\}$ is an RKH space, where $Q_{H0}(f, g) = \int_S$ <df, dM^{-1}dg> is defined in the obvious way. The reproducing kernel is K(s, t) = M(min(s, t)).

The following result is an easy extension of our scalar result.

Theorem 4.5. Let z in G_z be arbitrary and let $\{G_{H0}, Q_{H0}\}$ be the Hellinger RKH space generated by M(t) = Ez(t)z(t)*. Then $\{G_{H0}, Q_{H0}\}$ is the RKH space representation of z.

Having found the RKH space representation of z in G_z, we now turn our attention to finding the RKH space representation of h = Az. A simpleminded guess would be $H_{Az} = A(H_z) = \{f = Ag: g \in H_z\}$. It turns out that this guess is correct and we will verify it below.

Since we extended our definition of G_H to include matrix valued measures, we first need to similarly modify our definition of $\boldsymbol{A}$. The critical property to be preserved as we change $\boldsymbol{A}$ is for any $A \in \boldsymbol{A}$ to map G_H onto G_H. The most straightforward way to accomplish this is by requiring A to satisfy the estimate

$$|[Af](v) - f(v) - [Af](u) + f(u)| \leq c\int_u^v N_t(f)\, dm(t)$$

where the function m: S→R is chosen to be right continuous and nondecreasing and to satisfy $m(v) - m(u) \leq a/ \| M(v) - M(u) \|$ for some constant a. Here, $\| \cdot \|$ represents the operator norm induced by $| \cdot |$.

The property A: $G_H \to G_H$ holds if B in $\boldsymbol{B}$ maps G into G_H. The following calculation verifies B: $G \to G_H$, where $\boldsymbol{B}$ is defined with respect to the same function m as used for $\boldsymbol{A}$. Using the obvious notation, we have

$$\begin{aligned}\Sigma\ dBf^*dM^{-1}dBf &\leq \sum_{p=1}^{n} |\,[Bf](s_p) - [Bf](s_{p-1})\,|^2 \,\|\, (M(s_p) - M(s_{p-1}))^{-1} \,\| \\ &\leq c^2 N_T^2(f) \sum_{p=1}^{n} (m(s_p) - m(s_{p-1}))^2 \,\|\, (M(s_p) - M(s_{p-1}))^{-1} \,\| \\ &\leq c^2 a\, N_T^2(f)(m(t) - m(-r))\end{aligned}$$

where c is the constant corresponding to B in $\boldsymbol{B}$.

Since our modification of $\mathcal{A}$ only consisted of an extra restriction on m (which we previously called k), the algebraic properties of the new $\mathcal{A}$ and $\mathcal{B}$ follow from our results in Chapter I.

Exercise Find a suitable m if $M(t) = k(t)M$ for some constant matrix M and a right continuous, nondecreasing function $k: S \to R$. If all components of z are sums of Wiener processes, show that $dm(t) = dt$ is suitable.

We now find a representation of H_h for $h = Az$, with A in $\mathcal{A}$ and z in G_z. If f belongs to $A(H_z)$, then $f = A(A^{-1}f) = A(Ez(\cdot)c_f) = Eh(\cdot)c_f$ for some c_f in $L_2(z)$. Since $z = A^{-1}h$, c_f is also in $L_2(h)$, and hence f belongs to H_h. We have proved that $A(H_z)$ is contained in H_h. Rather than show H_h is contained in $A(H_z)$ directly (which is not very hard, but does not help identify an inner product on $A(H_z)$), for now we merely note that $R_h(\cdot, t)v$ belongs to $A(H_z)$ for each t in S and v in R^k. A logical candidate for an inner product on $A(H_z)$ is $<f, g>_A = <A^{-1}f, A^{-1}g>_z$. If f and g are in $A(H_z)$, we have $f = Eh(\cdot)c_f$ and $g = Eh(\cdot)c_g$ for some c_f, c_g in $L_2(z)$ and $L_2(h)$, and hence

$$<f, g>_A = <A^{-1}f, A^{-1}g>_z = <Ez(\cdot)c_f, Ez(\cdot)c_g>_z = Ec_f c_g$$
$$= <Eh(\cdot)c_f, Eh(\cdot)c_g>_h = <f, g>_h$$

Since $\{H_h, <\cdot, \cdot>_h\}$ is an RKH space, it follows that $\{A(H_z), <\cdot, \cdot>_A\}$ is an RKH space with the same reproducing kernel. The uniqueness of the RKH space representation of h gives us that the two spaces are identical.

Theorem 4.6. If A belongs to $\mathcal{A}$, z belongs to G_z, and $h = Az$, then $H_h = \{f = Ag: g \in H_z\}$ and $<x, y>_h = <A^{-1}x, A^{-1}y>_z$ for each x, y in H_h.

The above result gives us a tremendous amount of mathematical machinery with which to construct a formula for $R_h(s, t) = Eh(s)h(t)^*$. Since $AR_z(\cdot, s)v$ belongs to H_h and H_z for each s in S and v in R^d, we may use the reproducing properties of both H_h and H_z to compute $[AR_z(\cdot, s)](t)$. On the one hand, if v_1, v_2

belong to R^d, then

$$v_1^*[AR_z(\cdot, s)](t)v_2 = \langle R_h(\cdot, t)v_1, AR_z(\cdot, s)v_2\rangle_h$$
$$= \langle A^{-1}R_h(\cdot, t)v_1, R_z(\cdot, s)v_2\rangle_z = ([A^{-1}R_h(\cdot, t)](s)v_1)^*v_2$$

And on the other hand,

$$v_1^* [AR_z(\cdot, s)](t)v_2 = \langle R_z(\cdot, t)v_1, AR_z(\cdot, s)v_2\rangle_z$$
$$= \langle A^*R_z(\cdot, t)v_1, R_z(\cdot, s)v_2\rangle_z = ([A^*R_z(s, t)](s)v_1{}^*v_2$$

Since t in S and v_1, v_2 in R^d are arbitrary, we conclude that $A^{-1}R_h(\cdot, t) = A^*R_z(\cdot, t)$, and hence the following result holds.

Theorem 4.7. If A belongs to $\mathcal{A}$, z belongs to G_z, and $h = Az$, then $R_h(\cdot, t) = AA^*R_z(\cdot, t)$ for each t in S, where A^* is the adjoint of A with respect to $\{H_z, \langle\cdot, \cdot\rangle_z\}$.

It may appear at first glance that the above representation of R_h, while meeting our criterion of being in terms of A and R_z, will not be advantageous computationally because of the presence of A^*. We show below that R_h may be computed without knowing A^* explicitly, although it may in fact be more convenient to use the A^* formulation.

If s, t are in S and v_1, v_2 are in R^d, we have $\langle R_h(\cdot, s)v_1, R_h(\cdot, t)v_2\rangle_h = \langle AA^*R_z(\cdot, s)v_1, AA^*R_z(\cdot, s)v_2\rangle_z = \langle A^*R_z(\cdot, s)v_1, A^*R_z(\cdot, t)v_2\rangle_z$. Recalling that $A^*R_z(\cdot, s)$ is the matrix representation of A in $\{H_z, \langle\cdot, \cdot\rangle_z\}$, we may restate Theorem 4.7 in the following way. The RKH space representation of Az is given by the matrix representation of A with respect to the RKH space representation of z. The use of A^* may be avoided by noting that $v_1^*[A^*R_z(\cdot, t)](s)v_2 = \langle R_z(\cdot, s)v_1, A^*R_z(\cdot, t)v_2\rangle_z = \langle AR_z(\cdot, s)v_1, R_z(\cdot, t)v_2\rangle_z = [AR_z(\cdot, s)](t)v_1{}^*v_2$, from which it follows that $[A^*R_z(\cdot, t)](s) = [AR_z(\cdot, s)](t)^*$. To compute the variance of the ith component of the output h, it suffices to compute $N_H(A^*R_z(\cdot, t)e_i)^2$. The approximating sums for this norm contain only terms of the form $[AR_z(\cdot, s_p)](t)e_i$ for points s_p in the partition of S.

There are many applications of the above result. Since R_h and h are congruent, it should not be surprising that many systems properties of h may be stated in terms of R_h. We close this section by discussing a one-dimensional "filtering" problem for systems with incomplete observations.

For known operators A_1 in $\boldsymbol{A}$ and B_2 in $\boldsymbol{B}$, we assume that $h = A_1 z$ with z in G_z, and $y = B_2 h$. In this setting, it is traditional to think of h as the state of the system and y as the output. We measure the process y at discrete times $t_1, t_2, \cdots, t_n$, but are unable to measure or compute (since B_2 in $\boldsymbol{B}$ is not invertible) the state h at some time t_j. We wish to obtain the "best" estimate of $h(t_j)$ from the data $y(t_1), \cdots, y(t_n)$.

Specifically, we want to find the random variable h_j, sometimes denoted $E(h(t_j) \mid y(t_1), \cdots, y(t_n))$, such that h_j belongs to $L_2(y)$ and $|h(t_j) - h_j|_E = |h(t_j) - y|_E$ for all y in $L_2(y)$. That is, h_j is the orthogonal projection of $h(t_j)$ on $L_2(y)$. Since h_j is in $L_2(y)$, h_j is congruent to a function f_j in H_y. Then, our problem may be reformulated in a purely deterministic setting: find the projection f_j of $Ey(\bullet)h(t_j)$ on H_y. If we can find constants $c_1, \cdots, c_n$ such that $f_j = \Sigma_{i=1,n} c_i R_y(\bullet, t)$, then the congruence between H_y and $L_2(y)$ gives us $h_j = \Sigma_{i=1,n} c_i y(t_i)$.

We illustrate the method with a simple example. Suppose z is a standard Wiener process, $A_1 = I$, and $B_2 = 4C$. Then $y = 4Ch = 4Cz$ and, for $s \leq t$, $R_y(s, t) = 8(ts^2 - s^3/3.)$. Suppose we are given $y(1/4)$, $y(1/2)$, $y(3/4)$, and $y(1)$, and want to estimate $z(3/4)$. We search for c_1, c_2, c_3, and c_4 to minimize $E[(z(3/4) - \Sigma_{i=1,4} c_i y(t_i))^2]$. Equivalently, we want to write the projection of $Ey(\bullet)z(3/4)$ in terms of R_y. In this discrete-time setting, $Ey(\cdot)z(3/4)$ belongs to H_y, so we require a vector c such that $Rc = b$, where c_i are the components of c, $b_j = Ey(t_j)z(3/4)$, and $R_{ij} = R_y(t_i, t_j)$. The matrix R is invertible, and the solution is $c_1 = 24/97$, $c_2 = -84/97$, $c_3 = 273/1261$, $c_4 = 45/97$.

<u>Exercise</u>. A necessary condition for minimizing smooth $f(c_1, c_2, c_3, c_4)$ is that $\delta f/\delta c_i = 0$ for each i. Derive the equation $Rc = b$ by minimizing $f = E[(z(3/4) - \Sigma_{i=1,4} c_i y(t_i))^2]$.

<u>Exercise</u>. If $y = 4Cz$, then $z = y''/4$. Explain why a reasonable estimate of $z(3/4)$ is $(y(1) - y(1/2))/2$. For $c = (0, -1/2, 0, 1/2)^*$, compute b' and b.

4. Control of Systems with State Dependent Noise

In this section, we extend our notion of hereditary systems to include systems with state dependent noise. In particular, we will consider diffusion models of the form

$$h(t) = h(0) + \int_0^t b(s, h(s))\, ds + \int_0^t \sigma(s, h(s))\, dz(s) \tag{4.14}$$

where b and σ are linear in h, and z is in G_z. If σ depends on h, the random input $\int_{[0,t]} \sigma(s, h(s))\, dz(s)$ is a function of the state h, hence the term "state dependent noise". Equation (4.14) is often written $dh = bdt + \sigma dz$. However, z may be non-differentiable (e.g., z = Wiener process), so this equation is to be interpreted in the integral sense (4.14) rather than as a differential equation.

However, it is also true that the Wiener process W has unbounded variation in t on finite intervals for almost all ω. The last term in (4.14), then, cannot in general be understood as a Riemann-Stieltjes integral. Our first task below is to define integration with respect to a stochastic process.

<u>The belated stochastic integral</u>. There are many different types of stochastic integrals to be found in the literature. We will briefly discuss the two most prevalent, and contradictory, integrals before defining the particular integral we will use. Our goals in this discussion are to introduce concepts needed below as simply as possible, and to provide some historical perspective on our method.

To indicate the difficulties associated with stochastic integration, we first consider $\int_{[a,\, b]} W(t)\, dW(t)$, $0 \le a < b$, with W a standard Wiener process. A Rieman sum approximation of this integral would be $\sum_{p=1,\, N} W(u_p)\{W(s_p) - W(s_{p-1})\}$, $a = s_0 < s_1 < \cdots < s_N = b$. As shown in the exercise below, the limit of this sum as $N \to \infty$ depends on the values of u_p. If $u_p = s_{p-1}$, the limit (in the sense of convergence in $L_2(\Omega)$) is called the Ito integral of W with respect to W, and equals

$$(\mathrm{ITO})\int_a^b W(t)\, dW(t) = (1/2)[W^2(b) - W^2(a) - (b - a)] \tag{4.15}$$

The Stratonovich version of this integral is obtained with $u_p = (1/2)(s_{p-1} + s_p)$, and equals

$$(\mathrm{STRAT})\int_a^b W(t)\,dW(t) = (1/2)[W^2(b) - W^2(a)] \tag{4.16}$$

We are faced with a dilemma in defining a stochastic integral. While the Ito integral arises naturally in diffusion models and is an integral part of the powerful martingale theory, many values of Ito integrals do not coincide with standard calculus rules. For instance, the value (4.16) appears to be correct, while (4.15) looks incorrect.

A resolution offered by E. J. McShane [19], which we adopt in this chapter, is to use the Ito integral and follow modeling procedures, described below, which compensate for the unintuitive values of certain Ito integrals.

Exercise. Use Riemann sums to compute the expected values of the Ito and Stratonovich versions of $\int_{[a,\,b]} W(t)\,dW(t)$. Compare these with the expected values of the right-hand sides of (4.15) and (4.16).

The belated integral, as defined by McShane, differs only slightly from the Ito stochastic integral described above and from the left Stieltjes integral used in previous chapters. We define $\int_{[a,b]} f(t)\,dz(t)$, where f and z may both be random, as the limit of sums $\Sigma_{p=1,n} f(u_p)[z(s_p) - z(s_{p-1})]$ where $a = s_0 < \cdots < s_n = b$ and $a \leq u_p \leq s_{p-1}$ for $p = 1, \cdots, n$. The limit is taken with respect to the $L_{2,d}(\Omega)$ norm as the mesh, defined by $\max\{s_p - u_p\colon p = 1, \cdots, n\}$, of the partition tends to zero. The following result then holds.

Theorem 4.8. (McShane). Suppose that f in G and z in G are nonanticipating and there exists a positive constant K such that $|E((z(t) - z(s) \mid \boldsymbol{F}_s)| \leq K(t - s)$ and $E([z(t) - z(s)]^2 \mid \boldsymbol{F}_s) \leq K(t - s)$. Then $\int_{[a,b]} f(t)\,dz(t)$ exists.

Exercise. Verify that the Wiener process and mean zero Poisson process both satisfy the above hypotheses on z.

If z satisfies the hypotheses of Theorem 4.8, all of the terms in (4.14) are defined. Before investigating the existence and uniqueness of solutions of (4.14), we will discuss some modeling issues which will cause us to modify (4.14). The computation (4.15) should lead us to expect some surprises in stochastic integral calculus, and there are several unexpected results to be discovered. The basis of stochastic integral calculus, and the first of our surprises, is the Ito stochastic differential rule. For simplicity, we state the version which applies to the Wiener process.

Ito Stochastic Differential Rule. Suppose $b(t, x)$ has continuous partials b_t, b_x, and b_{xx}. Let x satisfy $dx = f(t)\,dt + g(t)\,dW(t)$, interpreted in the sense of (4.14). If $h(t) = b(t, x(t))$, then a.e. in Ω, $dh = b_t(t, x(t))\,dt + b_x(t, x(t))dx + (1/2)\, b_{xx}(t, x(t))\cdot g^2(t)\,dt$.

The term $(1/2)\, b_{xx}g^2dt$ in the expansion of dh does not appear in the analogous result from deterministic calculus, and is the source of "extra" terms appearing in computations like (4.15).

Exercise. Show $\int_{[a,b]} W(t)\,dW(t) = (1/2)\{W^2(b) - W^2(a) - b + a\}$. Hint: compute dW^2

Suppose for the moment that (4.14) has a unique solution with $z = W$ and that we want an algorithm for approximating this solution. Since W is not differentiable, (4.14) cannot be transformed into an ordinary differential equation. However, if we replace W with a piecewise linear approximation of W, we may solve the corresponding ordinary differential equation. If all goes well, these solutions will converge as our piecewise linear functions converge to W. The following result, which obviously relates to the Ito stochastic differential rule, indicates the unexpected flaw in this method.

Theorem 4.9. (Wong and Zakai [28]). Suppose that W is a Wiener process, and W_n is a piecewise linear approximation of W such that $W_n \to W$ a.e. as $n \to \infty$. Let x_n solve the ordinary differential equation

$$dx_n = b(t, x_n(t))\,dt + \sigma(t, x_n(t))\,dW_n(t) \tag{4.17}$$

Then x_n converges to the solution of

$$dx = b(t, x(t))\,dt + \sigma(t, x(t))\,dW(t) + (1/2)\sigma(t, x(t))\sigma_x(t, x(t))\,dt \qquad (4.18)$$

The fact that (4.17) converges to (4.18) rather than to (4.14) has dramatic implications in the simulation and modeling of stochastic systems. We will emphasize the modeling aspects of the result.

One view of the modeling process is that models should only be chosen from classes of mathematically "nice" models. Along with tractability, certain properties of consistency are necessary if useful information is to be extracted from the model. One such property relates to the sample paths of the noise process. The model noise (Wiener process) has unbounded variation a.e. in Ω, but it may be reasonable to presume that the physical noise process is Lipschitzian. A measure of consistency, and robustness, of the model is to require the model to be equally valid for Lipschitzian and non-Lipschitzian noise processes. Theorem 4.9 tells us that (4.14) is not consistent, since for non-Lipschitzian z the scheme (4.17) does work.

For the above reasons, we consider the model (4.14) to be inadequate. We will obtain a satisfactory model by extending (4.14) to a second order integral equation. In doing so, we will extend the operator spaces $\boldsymbol{A}$ and $\boldsymbol{B}$ and our notion of hereditary systems.

If z_1 and z_2 satisfy the hypotheses of Theorem 4.8 and $f \in G$ is nonanticipating, the second order belated integral $\int_{[a,b]} f(t)\,dz_1(t)\,dz_2(t)$ exists as the limit of sums $\sum_{p=1,n} f(u_p)\{z_1(s_p) - z_1(s_{p-1})\}\{z_2(s_p) - z_2(s_{p-1})\}$ where $a = s_0 < \cdots < s_n = b$ and $u_p \le s_{p-1}$. The second order integral simplifies in two important cases. If $\int_{[a,b]} f(t)\,dz_1(t)$ exists, then $\int_{[a,b]} f(t)\,dz_1(t)\,dz_2(t) = 0$. For the Wiener process W, $\int_{[a,b]} f(t)\,dW^2(t) = \int_{[a,b]} f(t)\,dt$ [191].

The above examples and Theorem 4.9 lead us to revise (4.14) to

$$h(t) = h(0) + \int_0^t b(s, h)\,ds + \int_0^t \sigma(s, h)\,dz(s) + (1/2)\int_0^t \sigma(s, h)\sigma_x(s, h)\,dz^2(s) \qquad (4.19)$$

In the case $z = W$, we note that (4.19) and (4.18) are equivalent a.e. If W is replaced by a piecewise linear approximation W_n, the second order integral in (4.19) vanishes, and (4.19) reduces to (4.17). By Theorem 4.9, (4.17) in turn converges to (4.19). On the other hand, if W is replaced by a Lipschitzian noise process, (4.19) reduces to (4.14) which is now solvable as an ordinary differentiable equation. Thus, (4.19) is the consistent version of (4.14).

Exercise. If z_1 and z_2 are differentiable and f in G is nonanticipating, show that $\int_{[a,b]} f(t)\, dz_1\, dz_2(t) = 0$.

Exercise. Compute the expected value of $\int_{[a,b]} f(t)\, dW^2(t)$.

We now return to the issue of existence and uniqueness of solutions, which we discuss for (4.19). Along with the hypotheses of Theorem 4.8, we assume that $E([z(t) - z(s)]^4 \mid \boldsymbol{F}_s) \leq K(t - s)$; both the Wiener process and Poisson process qualify. In this case, (4.19) has a solution. The solution is unique with the extra requirement that h satisfies (4.14) for all ω in Ω such that $z(\cdot, \omega)$ is Lipschitzian.

Extended hereditary systems. We now use the above theory of stochastic differential equations to extend the definition of hereditary systems to include systems with state dependent noise. We first rewrite the canonical stochastic model (4.19) as $h = f + B_1h + B_2h + B_3h$. Here, f contains all terms in (4.19) not involving h, and B_1, B_2, and B_3 are the integral operators corresponding to the ds, dW, and dW^2 terms, respectively. The following estimates relate these operators to $\boldsymbol{B}$ (note: we refer to $\boldsymbol{B}$ defined in section 1, rather than the matrix $\boldsymbol{B}$ introduced at the end of Section 3.) There exists a constant c such that for all nonanticipating f in G,

$$\left| \int_a^b f(t)\, dx(t) \right|^2 \leq c \int_a^b |f(t)|^2\, dt$$

$$\left| \int_a^b f(t)\, dz^2(t) \right| \leq c \int_a^b |f(t)|\, dt$$

Thus, B_3 is in $\boldsymbol{B}$. In general, B_2 is not in $\boldsymbol{B}$, but the above estimate shows us how to extend $\boldsymbol{B}$. A simple example of B_2 not in $\boldsymbol{B}$ is given by $[B_2f](t) = \int_{[0,t]} f(s)\, dW(s)$. If $f = 1$, we have that f belongs to G_H but $B_2f = W$ does not belong to G_H; therefore, B_2 is not an element of $\boldsymbol{B}$. To include operators of the type B_2 in our description of hereditary systems, we make the following definitions.

The opperator space $\boldsymbol{B}_E$ consist of all linear transformations B of G such that $P_0(Bf) = 0$ for all f in G, and whenever f belongs to G and $-r \leq u \leq v \leq T$,

$$| [Bf](v) - [Bf](u) |^2 \le c \int_u^v N_t^2(f)\, dk(t)$$

The operator space $\boldsymbol{A}_E$ consists of all operators I - B, where B belongs to $\boldsymbol{B}_E$.

An important feataure of the extensions $\boldsymbol{A}_E$ and $\boldsymbol{B}_E$ is that they retain the algebraic properties of $\boldsymbol{A}$ and $\boldsymbol{B}$.

Theorem 4.10. If A, A_1, A_2 are in $\boldsymbol{A}_E$ and B, B_1, B_2, are in $\boldsymbol{B}_E$, then the following hold.

1) I - B is 1 - 1 and onto G, and $(I - B)^{-1}$ belongs to $\boldsymbol{A}_E$;
2) A is 1 - 1 and onto G, A^{-1} is in $\boldsymbol{A}_E$, and $I - A^{-1}$ belongs to $\boldsymbol{B}_E$;
3) -B belongs to $\boldsymbol{B}_E$, $B_1 + B_2$ belongs to $\boldsymbol{B}_E$, and B_1B_2 belongs to $\boldsymbol{A}_E$;
4) $c_1A_1 + c_2A_2$ belongs to $\boldsymbol{A}_E$ whenever $c_1 + c_2 = 1$, A_1A_2 belongs to $\boldsymbol{A}_E$;
5) A - B belongs to $\boldsymbol{A}_E$, AB belongs to $\boldsymbol{B}_E$, and BA belongs to $\boldsymbol{B}_E$.

The proof of Theorem 4.10 resembles the proofs of Theorems 1.1 - 1.3. We sketch the proof of part (1) to indicate the differences in the details of the proofs.

We assume that B is in $\boldsymbol{B}_E$, f is in G, and c is the constant corresponding to B. Since $P_0B = 0$, it is sufficient to consider $S = [0, T]$. If $0 \le v \le T$, then $| [Bf](v) |^2 \le c\int_{[0,v]} N_t^2(f)\, dk(t)$. Applying this inequality twice, we get

$$| [B^2f[(v) |^2 \le c\int_0^v c\int_0^t N_s^2(t)\, dk(s)\, dk(t) = c^2 \int_0^v (k(v) - k(t))_t^2 (f)\, dk(t)$$

Induction yields

$$| [B^n f](v) |^2 \le c^n \int_0^v \{k(v) - k(t))^{n-1} N_t^2(f)/(n-1)!\}\, dk(t)$$

It follows that $| [B^n f](v) |^2 \le a_n N_t(f)^2$, where $a_n = c^n(k(t) - k(0))^n/(n-1)!$. We call on the Ratio Test to conclude that $\Sigma_{n=1,\infty} | [B^n f](v) |$ converges. Since $\Sigma_{n=0,\infty} B^n f$ converges everywhere in S, $h = f + Bh$ has a unique solution. This establishes that I - B is one to one and onto, and hence has an inverse.

We name $(I - B)^{-1} = D$, so that $D - I = BD$. To show that D belongs to $\boldsymbol{A}_E$, we show that BD belongs to $\boldsymbol{B}_E$. To that end we note that

$$| [BDf](t) |^2 \le c \int_0^t N_s^2(Df)\, dk(s) \le 2c \int_0^t \{N_s^2(f) + N_s^2((D - I)f)\}\, dk(s) \qquad (4.20)$$

Since D - I = BD, we may replace the second term in (4.20) with the inequality given in (4.20). After integrating by parts, we repeat the above procedure to obtain

$$| [BDf](t) |^2 \leq 2c \exp\{2c(k(t) - k(0))\}\int_0^t N_s^2(t)\, dk(s) \tag{4.21}$$

We use (4.21) in the second line below. A series of steps similar to those used to derive (4.21) produce the following proof that BD belongs to $\boldsymbol{B}_E$:

$$\begin{aligned}
|[BDf](v) - [BDf](u)|^2 &\leq 2c\int_u^v \{N_s^2(f) + N_s^2((D - I)f)\}\, dk(s) \\
&\leq 2c\int_u^v \{N_s^2(f) + 2c\exp(2c(k(T) - k(0)))\int_0^s N_w^2(f)\, dk(w)\}\, dk(s) \\
&\leq 2c[1 + 2c\exp(2c\{k(T) - k(0)\})](k(T) - k(0))\int_u^v N_t^2(f)\, dk(t)
\end{aligned}$$

Thus, D is in $\boldsymbol{A}_E$, completing the proof of Theorem 4.10, part (1). □

Exercise. Prove parts (ii) - (v) of Theorem 4.10.

Control of systems with state dependent noise. The significance of Theorem 4.10 is that we may expect to analyze systems of the form (4.19) with essentially the same methods we used in Section 3. As with the generalization of state space systems to hereditary systems, the price we will have to pay for allowing more complicated dynamics involves the cost functional to be minimized.

In Section 3, the penalty assigned to the control of the system was of the form $N_H^2(Au)$. The following example shows that this term is not appropriate for systems with state dependent noise. Suppose that $[Bf](t) = \int_{[0,t]} f(s)\, dW(s)$ and Then $Af = 1 - W$ does not belong to G_H. Thus for B in $\boldsymbol{B}_E$, we are not guaranteed that A in $\boldsymbol{A}_E$ maps G_H onto G_H.

Our method is to first solve a deterministic problem for which our standard cost functional is defined, and then solve the related stochastic problem as a corollary. Instead of the extended stochastic hereditary system dynamics $h = B_1h + B_2h + f + z + u$ with B_1 in $\boldsymbol{B}$ and B_2 in $\boldsymbol{B}_E$ we consider the deterministic system $h = B_1h + f + u$ with B_1 in $\boldsymbol{B}$. We note that if B_2h consists of terms of the form $\int_{[0,t]} m(t, s)h(s)\, dz(s)$, $EB_2h = 0$. The control problem is to find $u = Dh$ with D in

B, to minimize

$$J(h, u) = (1/2)\, N_H^2(Ch) + (1/2)b\, N_H^2(A_1u)$$

where b is a constant, subject to $h = A_1(f + u)$ and $h = 0$ on $[T - r, T]$. Instead of assigning a penalty to nonzero terminal values of the output h, we insist that h be zero on an interval of length r. We recall that r is at least as large as any delays in the system. Thus, we force the system to be truly at rest, rather than just requiring the output to hit 0 at some specified time. The nonzero (deterministic) input f must be overcome to achieve this.

As always, we apply the Lagrange Multiplier Theorem and Lemma 4.1 to obtain a necessary condition. To reduce the notation, we write A for A_1, and set $\alpha = 1/b$. The constraint $h = 0$ on $[T - r, T]$ is enforced as separate constraints $h(T) = 0$ and $(I - P_{T-r})h = 0$.

For (h, u) to minimize J subject to the above conditions, there must exist λ in G_H, μ in R, and γ in G_H such that $P_{T-r}\gamma = 0$ and

$$Q_H(Ch', Ch) + bQ_H(Au', Au) + Q_H(h' - Au', \lambda) + Q_H(h', \mu k) + Q_H(h', \gamma) = 0 \quad (4.22)$$

for all u′, h′ in G_H.

Setting $h' = 0$ in (4.22) and requiring $P_0u = 0$ gives us $bAu = (I - P_0)\lambda$. To find λ, we set $u' = 0$ in (4.22) and apply Lemma 4.2 to obtain $C^*Ch + \lambda + \mu k + \gamma = 0$. We solve for μ by evaluating this expression at $t = 0$. With μ and our representation of C*, we find $\lambda = C^2h + k(\lambda/k)(0) - \gamma$. That is,

$$bAu = C^2h + (\lambda/k)(0)(I - P_0)k - \gamma \quad (4.23)$$

We use the three constraints to solve for $\lambda(0)$ and γ.

The first constraint is $h = A(f + u)$. We substitute (4.23) and require $h(T - r) = 0$. Then

$$0 = [(I - \alpha C^2)^{-1}Af](T - r) + \alpha[(I - \alpha C^2)^{-1}(I - P_0)k/k(0)](T - r)\lambda(0) \quad (4.24)$$

which identifies $\lambda(0)$. Combining $h = A(f + u)$, (4.23), and the remaining constraint $(I - P_{T-r})h = 0$ gives us

$$\gamma = (I - \alpha C^2)(I - P_{T-r})(I - \alpha C^2)^{-1}(Af + \alpha(I - P_0)k(\lambda/k(0)) \qquad (4.25)$$

The causality of elements of $\boldsymbol{A}$ and $\boldsymbol{B}$ guarantees $\gamma = (I - P_{T-r})\gamma$, as desired. The necessary conditions (4.23) - (4.25) may be shown to be sufficient in the usual manner. We have the following result.

Theorem 4.11. Let Γ be all deterministic pairs (h, u) in $G \times G_H$ with $P_0 u = 0$. Define J on Γ by $J(h, u) = (1/2)N_H(Ch)^2 + (1/2)bN_H(Au)^2$. Then J achieves a minimum on Γ subject to $h = A(f + u)$ and $h = 0$ on $[T - r, T]$ with $u = \alpha A^{-1}\{C^2h + (\lambda/k(0))\,(I - P_0)\,k - \gamma\}$ where $\alpha = 1/b$ and $\lambda(0)$ and γ are given by (4.24) and (4.25), respectively.

We now return to the stochastic system $h = B_1h + B_2h + f + u + z$. If f is a non zero deterministic input, the first goal of a control function would be to control the mean of the output. This is accomplished by applying Theorem 4.11 in the following way.

Theorem 4.12. Suppose f belongs to G, $Ef = f$, z belongs to G_0, B_1 belongs to B, $\boldsymbol{B}_2$ belongs to $\boldsymbol{B}_E$, $EB_2 = 0$, $A = (I - B_1 - B_2)^{-1}$, and $A_1 = (I - B_1)$. Define $J(h, u)$ on $G \times G_H$ by $J(h, u) = (1/2)N_H(CEh)^2 + (1/2)\,b\,N_H(A_1Eu)^2$. Then J achieves a minimum on $G \times G_H$ subject to $h = A(f + u + z)$ and $Eh = 0$ on $[T - r, T]$ with $u = \alpha A_1^{-1}\{C^2h + (\lambda/k(0))(I - P_0)k - \gamma\}$ where $\alpha = (1/b)$ and $\lambda(0)$ and γ are given by (4.24) and (4.25), respectively.

Under the hypotheses of Theorem 4.12, Eh and Eu satisfy the hypotheses of Theorem 4.11. Furthermore, the optimal control of Theorem 4.11 satisfies $A_1Eu = EAu$, so theorem 4.10 follows.

We note some properties of the optimal control found in Theorem 4.12. Since f determines $\lambda(0)$ and γ, u consists of a simple feedback term $\alpha(I - B_1)C^2h$, and an open loop term which may be computed in advance. Furthermore, γ is only nonzero on the interval $[T - r, T]$.

To further explore the nature of the optimal control, we return to the pendulum problem discussed in Section 3. The displacement angle $\theta(t)$ of the pendulum satisfies $\theta'' + \theta' + \theta = w$, $\theta(0) = a$, $\theta'(0) = b$. We have previously derived 1-, 2-, and 3-dimensional models, all of which require $E(a) = E(b) = 0$. Under this assumption, the control obtained from Theorem 4.12 is identical to the control obtained from Theorem 4.4.

The assumption $E(a) = E(b) = 0$ indicates the following interpretation of the control: the pendulum starts in the zero state, and the control is designed to keep the pendulum as close to zero as possible. If $E(a)$ or $E(b)$ is nonzero, the control must drive the pendulum from its initial state to the zero state in the most cost efficient manner. In the discussion to follow, we assume $E(a) = x$ and $E(b) = y$, where not both of the real constants x and y are zero.

The hereditary system 1-dimensional model is obtained by integrating (4.7) twice. We identify $f(t) = x + (x + y)t$, $B_3 = -C - C^2$, and $z_1 = C^2w$. In Section 3, we assumed that the disturbance w was state independent. Here, we assume that $[C^2z](t) = z(t) + \int_{[0,t]} \theta(s)\, dW(s)$, where W is a Wiener process and z belongs to G_0. The model then becomes $\theta = f + z + B_3\theta + B_2\theta$.

This model, however, is of the form (4.14). The appropriate model of the form (4.19) is $\theta = f + z + B_1\theta + B_2\theta$, where $B_1 = B_2 + (1/2)C$. To compute the optimal control of Theorem 4.12, it is first necessary to find (4.24) and (4.25). If $b = 1$, we first compute $[(I - C^2)^{-1}(I - P_0)k](t) = \sinh(t)$. The other function of interest is $[(I - C^2)^{-1}A_1f](t)$, where $A_1 = (I - B_1)^{-1}$. Since $B_1 = C^2 - (C/2)$, we may use Laplace transforms to compute

$$[A_1f](t) = f(t) + \int_0^t g(t-s)f(s)\, ds$$

$$g(t) = -1/2\{\cos(mt/4)+7\sin(mt/4)/m\}\exp(-t/4)$$

$$m^2 = 15$$

$$[(I - C^2)^{-1} A_1f](t) = [A_1f](t) + \int_0^t \sinh(t-s)[A_1f](s)\, ds$$

Using $f(t) = x + (x + y)t$, we may write $[(I - C^2)^{-1}A_1f](t)$ in the form

$$x\cosh(t) + (x + y)\sinh(t) + [b_1 + b_2\cos(mt/4) + b_3\sin(mt/4)]\exp(-t/4)$$

where b_1, b_2, and b_3 are constants depending on x and y.

Thus, the open loop control term depends only on elementary functions and the values x and y of $\theta(0)$ and $\theta'(0)$, respectively.

Summary

In the first sections of this chapter, we presented the definitions and probabiistic concepts necessary to solve the control problems in the chapter. Assuming a basic knowledge of measure theory, we developed many properties of second order stochastic processes. Billingsley [4] is a good source for the prerequisite measure theory, with particular emphasis on probability measures. Many of our biases towards second order processes derive from Grenander [12], who delves into the area of statistical inference which we briefly touch upon in Section 3. An excellent reference for any of the topics introduced in the first two sections is Feller [10].

An alternative approach to stochastic processes is martingale theory. Elliott [8] gives an introduction to this theory, with a clear treatment of the stochastic calculus discussed in Section 4 here. Wong and Hajek [27] apply these methods to many of the topics discussed in this chapter. A nice, and unusual in the literature, feature of [27] is that the main results are valid for both continuous and jump noise processes. We believe that this feature, which holds for all of our results, is one of the most important aspects of our work.

With all our machinery in place, we opened Section 3 by solving a control problem for linear systems. The problem we solve is very similar to the classical linear regulator or linear-quadratic-gaussian control problems which are much discussed in the literature. Leitmann [15] gives a geometric derivation of the maximum principle, which is basic to the classical techniques for solving linear control problems. Fleming and Rishel [11] derive necessary conditions for the optimal control of the linear-quadratic-gaussian problem using the dynamic programming approach. The optimal control is obtained by solving a partial differential equation. However, in the state space case, the p.d.e. reduces to a matrix Riccati ordinary differential equation. This Riccati equation is obtained by Russell [24] using calculus of variations and matrix theory. We present Russell's results to give a more detailed comparison of methods.

We write the system dynamics as

$$dx = [A(t)x(t) + B(t)u(t)]\,dt + \sigma(t)\,dW(t)$$

where W is a Wiener process. The cost functional to be minimized is

$$J = E\int_0^T \{x^*(t)M(t)x(t) + u^*(t)N(t)u(t)\}dt + Ex^*(T)Dx(T)$$

for positive definite matrices M, N and D. The J we consider in Section 3 is equivalent to the above with $M(t) = (1/2)c_1I$, $N(t) = (1/2)c_2I$, and $D = (1/2)I$. The optimal control is required to have the form $u(t) = K(t)x(t)$, where K(t) is called the gain matrix.

Under the above conditions, the optimal control is given by $u(t) = -N(t)^{-1}B^*(t)\cdot Q(t)x(t)$, where Q satisfies the Riccati equation

$$Q'(t) = -Q(t)A(t) - A^*(t)Q(t) + Q(t)B(t)N(t)^{-1}B^*(t)Q(t) - M(t) \qquad Q(T) = D$$

To implement the control then, it is necessary to have the solution of the above matrix o.d.e. for all t.

In contrast, we allow the control to have the form $u = Dh$, $D \in \boldsymbol{B}$. Certainly, if $u'(t) = K(t)h(t)$, we recall that our control is added to the integral equation $h = A(u + z)$ rather than to the above differential equation, then $u = Dh$ with $D = CK$. Thus, our space of admissible controls is larger, and we are able to find an explicit formula for the optimal feedback operator D.

A further comparison may be made by differentiating the optimal (Russell) control $u = Kx$ and substituting for K' and h'. After integrating twice, it can be shown that $u_0 = C_u$ satisfies

$$u_0 = (I + \alpha^*C)^{-1}C^2h + (I + \alpha^*C)^{-1}C^2Kz'$$

The difference between u_0 and our optimal control is the second term in the above expression.

In contrast to the various techniques, loosely described as direct computation and RKH space methods, for computing covariance kernels in Section 3 here, Russell derives a Riccati equation for the covariance. If $X_0 = Ex_0x_0^*$, the covariance $X(t) = Ex(t)x(t)^2$ satisfies

$$X'(t) = S(t)X(t) + X(t)S(t)^* + C(t)V(t)C(t)^* \qquad X(t_0) = X_0$$

where $V(t) = Ev(t)v(t)^*$, $v = W'$, and $S = A + BK$. We note that if x does not have independent increments, $\{X(t) : t \in [0, T]\}$ does not determine $R_X(s, t)$ for s, t in [0, T].

Curtain and Pritchard [5] are able to derive similar results for distributed parameter systems, which include delay systems and dynamics described by partial differential equations. The formulation for the delay system control problem (see [3], for example) implies that the base space X is infinite dimensional. The resulting complications, infinite-dimensional noise and infinite dimensional matrix differential equations, make actual computations formidable. Several techniques for computing finite dimensional approximations (see [22], for example) have been used.

Our method avoids the above problems for delay systems. Our only concession to the advanced diffuculty of the problem is the change in cost functional, to a nonstandard but intuitively reasonable cost, discussed in the preamble to Theorem 4.4.

The material in Section 3 on RKH space methods in time series analysis was introductory. The benchmark papers in this area are collected in [25]. As noted in the filtering discussion, the ability to compute a covariance kernel explicitly has consequences far beyond measuring the size of a process. For example, Duttweiler and Kailath [6] solve a parameter estimation problems by factoring the operators involved.

It is obvious that our treatment of systems with state dependent noise owes a lot to the work of McShane [19]. The reader is urged to read [19] for more detailed descriptions of the properties of belated integrals and extended hereditary systems. There are several important references which are less related to our work. Karlin and Taylor [14] have an excellent chapter on diffusion processes. Arnold [1] discusses stochastic differential equations.

In Section 4, we defined extended hereditary systems to include systems with state dependent noise. In Theorem 4.12, we solved a control problem for extended hereditary systems, using essentially the same techniques as in Section 3. The complexity of the problem forced us to redefine the cost functional in terms of expected values of the output and control.

The benchmark paper on control of systems with state dependent noise is by Wonham [29]. Most of the literature focuses on bilinear systems (see [13], for example), rather than on the linear systems considered here.

In summary, the general hereditary system framework has been used to solve control problems for stochastic systems with state space, delay, and state dependent noise dynamics. Furthermore, the RKH space methods of time series analysis are readily applicable for problems of state estimation and parameter estimation.

The following bibliography includes all papers referenced in this chapter, as well as a small number of other important references. We do not present an exhaustive list (an excellent bibliography of that type may be found in [11]), but attempt to give at least one outside reference for each of the topics alluded to in the chapter.

References

1. L. Arnold, Stochastic Differential Equations, John Wiley, New York, 1974.

2. N. Aronszajn, Theory of reproducing kernels, Trans. AMS, 68(1950), 337-404.

3. H. T. Banks and J. A. Burns, An abstract framework for approximate solutions to optimal control problems governed by hereditary systems, Proc. Int. Conf. on Differential Equations, H. A. Antosiewicz ed., Academic Press, New York, 1975, 10-25.

4. P. Billingsley, Probability and Measure, John Wiley, New York, 1979.

5. R. F. Curtain and A. J. Pritchard, Infinite Dimensional Linear Systems Theory, Springer-Verlag, Berlin, 1978.

6. D. L. Duttweiler and T. Kailath, RKHS approach to detection and estimation problems-part 5: parameter estimation, IEEE Trans. Inform. Theory 19(1973), 29-37.

7. A. Einstein, Investigations on the Theory of the Brownian Movement, Ed. R. Furth, Dover, New York, 1956.

8. R. J. Elliott, Stochastic Calculus and Applications, Springer-Verlag, New York, 1982.

9. A. Feintuck and F. Saeks, Systems Theory: A Hilbert Space Approach, Academic Press, New York, 1982.

10. W. Feller, An Introduction to Probability Theory and Its Applications,Volume 1, third edition, John Wiley, New York, 1968.

11. W. H. Fleming and R. W. Rishel, Deterministic and Stochastic Optimal Control, Springer-Verlag, New York, 1975.

12. U. Grenander, Abstract Inference, Wiley-Interscience, New York, 1981.

13. A. Ichikawa, Dynamic programming approach to stochastic evolution equations, SIAM J. Control and Optim. 17(1979), 152-174.

14. S. Karlin and H. M. Taylor, A First Course in Stochastic Processes, second edition, Academic Press, New York, 1975.

15. G. Leitmann, The Calculus of Variations and Optimal Control, Plenum Press, New York, 1981.

16. C. C. Lin and L. A. Segel, Mathematics Applied to Deterministic Problems in the Natural Sciences, MacMillan, New York, 1974.

17. J. S. Mac Nerney, Hellinger integrals in inner product spaces, J. Elisha Mitchell Scientific Society 76 (1960), 252-273.

18. B. B. Mandlebrot, The Fractal Geometry of Nature, W. H. Freeman and Company, New York, 1983.

19. E. J. McShane, Stochastic Calculus and Stochastic Models, Academic Press, New York, 1974.

20. R. B. Minton and J. A. Reneke, An RKH space approach to state feedback control of a class of linear stochastic systems, SIAM J. Control and Optim., 24(1986), 700-714.

21. E. Parzen, An approach to time series analysis, Am. Math. Stat. 32(1961), 951-989.

22. R. K. Powers, Chandrasekhar equations for distributed parameter systems, dissertation, Virginia Polytechnic Institute and State University, 1984.

23. J. A. Reneke and R. B. Minton, RKH space methods for approximating the covariance kernels of a class of stochastic hereditary systems, II, Frequency Domain and State Space Methods for Linear Systems, C. J. Byrnes and A. Lindquist, Eds., North-Holland, Amsterdam, 1986, 423-429.

24. D. L. Russell, Mathematics of Finite Dimensional Control Systems: Theory and Design, Marcel Dekker, New York, 1979.

25. H. L. Weinert, ed., Reproducing Kernel Hilbert Spaces, Hutchinson Ross Publishing Company, 1982.

26. N. Wiener, The average of an analytical functional and the Brownian movement, Proc. Nat. Acad. of Science 7(1921), 294-320.

27. E. Wong and B. Hajek, Stochastic Processes in Engineering Systems, Springer-Verlag, New York, 1985.

28. E. Wong and M. Zakai, On the relation between ordinary and stochastic differential equations, Int. J. Engineering Sciences 3(1965), 213-229.

29. W. M. Wonham, Optimal stationary control of a linear system with state-dependent noise, SIAM J. Control and Optim. 5(1967), 486-513.

V
Large Scale Systems

1. Introduction

Large scale systems are made up of components which can exert some level of autonomous control. A simple example would be a house with a thermostat in each room, i.e., a sensor and control for a heat source available to each room. We can get at the difficulty of controlling a large scale system by imagining the following situation for the house with multiple thermostats. The objective is to raise the household temperature from one equilibrium to a higher one.

In order to have interaction between the components, the rooms should not be completely insulated allowing heat to flow from one to another. Suppose we set each thermostat at the new temperature. If one room, probably the smallest, reaches the new temperature before the others, its heat source is cut off only to be switched back on after a heat loss to the large rooms. This switching off and on could be repeated by several of the rooms until the slowest to respond finally reaches the new temperature.

The situation becomes even more complicated if the individual controllers attempt to take advantage of the flow of heat from a warmer to a colder room. The difficulties of this example are intrinsic to large scale systems even when we assume cooperation between the controllers. For instance, if the controllers exchange information on what actions they plan to take, what each does is influenced by the actions of the others, a seemingly vicious circle.

The problem of controlling the temperature for such a large scale system might be contrasted with the control problem for a house with a sensor in each room and a single controller. The implication being that the controller in the latter case bases his decision for supplying heat to any room on the temperatures of all the rooms. The single controller can operate the multiple heat sources to achieve some overall optimum.

Our objective in the design of control laws for large scale systems is the effective coordination of the interacting components each of which has some level of autonomous control. Coordination is possible when the interactions between components is suitably limited, i.e., the components are "stably connected". Coordination of the components is achieved through the implementation of a control strategy consisting of a set of side conditions imposed on the components together with the control laws that result from local optimization. The side conditions, conditions added to the system objectives, are in the form of interval conditions on the component responses. Choice of appropriate "coordination parameters" which appear in these side conditions leads to a higher level coordination problem.

Chapter Outline. In general, we are interested in choosing controls which steer a large scale system from one position to another or which force the system variables to follow some desired trajectory. We begin, in Section 2, by considering two local optimization problems with the type of side conditions required for the large scale problem. Stably connected two component systems are introduced in Section 3 and a control coordination strategy is developed for such systems. This coordination strategy is first applied to control problems for finite dimensional systems using an appropriate decomposition. Two state space examples are explored illustrating concretely the concept of system coordination. We also consider an hereditary system abstraction of the household heating problem. We conclude this section by extending our ideas for two component systems to more general network systems. The chapter concludes with a discussion of higher level coordination problems, possibilities that arise in the absence of centralized control.

2. The Local Optimization Problem

The control coordination strategy to be introduced requires that component interactions be suitably limited and certain side conditions be added to the component objectives. Local control problems can then be solved without full information of the actions of the other component controllers. In this section we consider two optimization problems, each is typical of the type of local optimization problem which must be solved by component controllers. Lagrange multiplier methods will be used to characterize solutions of the component control problems.

We are concerned with system components whose signal diagram is of the form

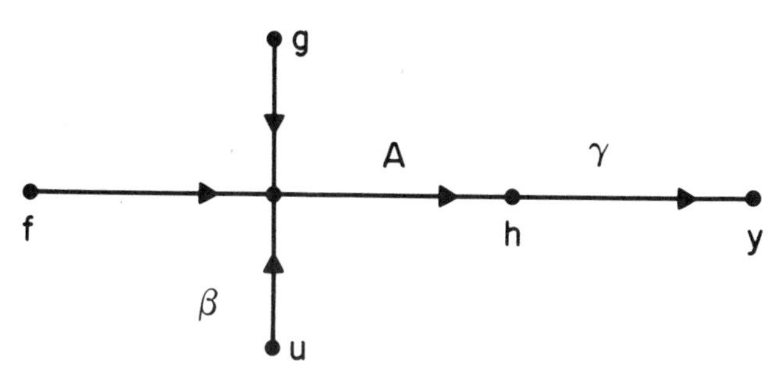

Figure 5.1

Here $A = (I - B)^{-1}$ with B in $\boldsymbol{B}$; f in G_H is a known input signal; g in G_H is an unknown input; u in G_H is the component control; and ß and γ denote linear transformations from X to X. The signal g represents the effect of other components upon the given component. We assume $P_0 g = 0$ and will add other conditions necessary for the solution of the local optimization problems.

Returning to the setting of Chapter II, let $S = [-r, T]$, $[Ch](t) = 0$ for $-r \leq t \leq 0$, and $[Ch](t) = \int_{[0,t]} h(\tau)\, d\tau$ for $0 \leq t \leq T$. For the rest of this section, assume that B is a fixed member of $\boldsymbol{B}$ and ß and γ are fixed $d \times d$ matrices. Let U be a subspace of X and Ω denote the set of u in G_H with the property that $P_0 u = 0$ and u(t) belongs to U for each t in S. We will be concerned with control systems $h = f + Bh + ßu$ or, alternately, $h = A(f + ßu)$, where $A = (I - B)^{-1}$, f and h are in G_H, and the control u is in Ω.

We will say the system (A, ß) or, alternately, (B, ß) is <u>controllable</u> provided $Aß = (I - B)^{-1}ß$ maps Ω onto $(I - P_0)G_H$, where $U = \boldsymbol{R}(ß^*)$. Assume for the rest of this section that (A, ß) is controllable. Note that for $ß = I$ the invertibility of the operator A guarantees controllability. If $B = \alpha C$, where α is a $d \times d$ matrix, then the condition is equivalent to rank $[\beta, \alpha\beta, \alpha^2\beta, \ldots, \alpha^{d-1}\beta] = d$. For delay differential equations (Example 2.1), this assumption implies function space controllability [1, 7, 14].

As a first example, we want to solve the following optimal control problem. Let D be in ***B*** and b_1 and b_2 be in X.

<u>Local Optimization Problem I</u> (LOP I)

Minimize on $\Omega \times G_H$: $J(u, h) = (1/2)N_H(A\beta u)^2 + (1/2)N_H(Ch)^2$

Subject to:

$$h = A(f + g + \beta u),$$
$$h(t) = 0 \text{ for } T\text{-}r \leq t \leq T,$$
$$[Dh](T - r) = b_1,$$

where g in G_H is unknown but fixed, $P_0 g = 0$, $[(I - C^2)^{-1}Ag](T - r) = b_2$, and $[D(1 - C^2)^{-1}Ag](T - r) = 0$.

Here the control objective is to force the response h to zero over the terminal interval. The conditions $[(I - C^2)^{-1}Ag](T - r) = b_2$ and $[D(1 - C^2)^{-1}Ag](T - r) = 0$ summarize information about g that allows the controller to solve the optimization problem. The rationale for these restrictions will become apparent in the proof. In the performance index for this example the term $(1/2)N_H(Ch)^2$ corresponds to the standard quadratic cost functional $(1/2)\int_{[0,T]} h^2(t)\,dt$. The term $(1/2)N_H(A\beta u)^2$ represents a penalty on the response of the system rather than a direct penalty on the control energy. It should be noted that the condition $h(t) = 0$ for $T - r \leq t \leq T$ is equivalent to the conditions $h(T) = 0$ and $(I - P_{T-r})h = 0$.

The Lagrange Multiplier Theorem asserts that if (u, h) minimizes J subject to the constraints, then there exists an element $(\kappa, \lambda, \mu, \upsilon)$ of $(I - P_{T-r})G_H \times G_H \times X \times X$ such that

$$Q_H(A\beta u', A\beta u) + Q_H(h', C^*Ch) + Q_H(h', \kappa)$$
$$+ Q_H(h' - A\beta u', \lambda) + Q_H(h', K(\;, T)\mu) + Q_H(h', D^*K(\;, t - r)\upsilon) = 0$$

for all (u', h') in $\Omega \times G_H$.

Hence, $A\beta u = (I - P_0)\lambda$ and $C^*Ch + \kappa + \lambda + K(\;, T)\mu + D^*K(\;,T - r)\upsilon = 0$. Eliminating μ, we have

$$C^*Ch + \kappa + \lambda - (1/k(0))\, k[C^*Ch](0) - (1/k(0))\, k\lambda(0)$$
$$- (1/k(0))\, k[D^*K(\ , T - r)](0)\upsilon + D^*K(\ , T - r)\upsilon = 0$$

or, since $C^*Ch = (1/k(0))k[C^*Ch](0) - C^2h$,

$$\lambda = C^2h + (1/k(0))\, k\lambda(0) + (1/k(0))\, k[D^*K(\ , T - r)](0)\upsilon - D^*K(\ , T - r)\upsilon - \kappa \tag{5.1}$$

Since $h = A(f + g) + (I - P_0)\lambda$, using (5.1) we have

$$h = A(f + g) + C^2h + k_1\lambda(0) + k_2\upsilon - \kappa = (I - C^2)^{-1}\,\{A(f + g) + k_1\lambda(0) + k_2\upsilon - \kappa\} \tag{5.2}$$

where $k_1 = (1/k(0))(I - P_0)k$ and $k_2 = (1/k(0))(I - P_0)k[D^*K(\ , T - r)](0) - (I - P_0)\cdot D^*K(\ , T - r)$. From (5.2) and the observation that $[(I - C^2)^{-1}\kappa](T - r) = 0$ and $[D(I - C^2)^{-1}\kappa](T - r) = 0$, we obtain the following linear system of equations for $\lambda(0)$ and υ:

$$\begin{aligned} 0 &= [(I - C^2)^{-1}Af](T - r) + b_2 + [(I - C^2)^{-1}k_1](T - r)\lambda(0) + [(I - C^2)^{-1}k_2](T - r)\upsilon \\ b_1 &= [D(I - C^2)^{-1}Af](T - r) + [D(I - C^2)^{-1}k_1](T - r)\lambda(0) + [D(I - C^2)^{-1}k_2](T - r)\upsilon \end{aligned} \tag{5.3}$$

Since we started with the assumption that (u,h) minimizes J, the system of equations (5.3) has a solution, perhaps not unique. Let $(\lambda(0), \upsilon)$ be any solution,

$$\kappa = (I - C^2)(I - P_{T-r})(I - C^2)^{-1}\,\{A(f + g) + k_1\lambda(0) + k_2\upsilon\} \tag{5.4}$$

$$h = (I - C^2)^{-1}\{A(f + g) + k_1\lambda(0) + k_2\upsilon - \kappa\} \tag{5.5}$$

and

$$u = (A\beta)^{\dagger}(C^2h + k_1\lambda(0) + k_2\upsilon - \kappa) \tag{5.6}$$

where $(A\beta)^{\dagger}$ denotes the pseudoinverse of $A\beta$.

Note that $(Aß)^\dagger = ß^\dagger(I - B)$, i.e., u has values in $R(ß^*)$, $P_0u = 0$, and $(I - P_{T-r})h = 0$. Since $Aß\Omega = (I - P_0)G_H$, $Aß(Aß)^\dagger w = w$, for each w in $(I - P_0)G_H$. Hence $A(f + g + ßu) = A(f + g) + C^2h + \kappa_1\lambda(0) + k_2\upsilon - \kappa = h$. Finally, $[Dh](T - r) = [D(I - C^2)^{-1}Af](T - r) + [D(I - C^2)^{-1}k_1](T - r)\lambda(0) + [D(I - C^2)k_2](T - r)\upsilon = b_1$, i.e., (u, h) defined by (5.4-6) is a feasible solution.

If (u′, h′) is a feasible solution then

$$
\begin{aligned}
&Q_H(Aßu', Aßu) + Q_H(Ch', Ch)\\
&\quad = Q_H(h' - A(f + g), Aßu) + Q_H(h', C^*Ch)\\
&\quad = Q_H(h', C^*Ch + Aßu) - Q_H(A(f + g), Aßu)\\
&\quad = Q_H(h', -C^2h + Aßu) - Q_H(A(f + g), Aßu)\\
&\quad = Q_H(h', -D^*K(\ , T - r)\upsilon) - Q_H(P_0f, \lambda) - Q_H(A(f + g), Aßu)\\
&\quad = -\langle b_1, \upsilon\rangle - Q_H(P_0f, \lambda) - Q_H(A(f + g), Aßu)\\
&\quad = Q_H(Aßu, Aßu) + Q_H(Ch, Ch)
\end{aligned}
$$

Thus

$$
\begin{aligned}
0 &\le N_H(Aß(u' - u))^2 + N_H(C(h' - h))^2\\
&= N_H(Aßu')^2 + N_H(Ch')^2 - N_H(Aßu)^2 - N_H(Ch)^2\\
&= 2J(u', h') - 2J(u, h)
\end{aligned}
$$

i.e., $J(u, h) \le J(u', h')$ or (u, h) minimizes J. This argument also shows that if (u, h) and (u′, h′) minimize J subject to the constraints then $ßu = ßu'$ and $h = h'$.

Therefore if (5.3) has a solution then LOP I has a solution. This motivates us say that LOP I is <u>regular</u> provided

$$
\operatorname{rank}\begin{bmatrix} [(I - C^2)^{-1}k_1](T - r) & [(I - C^2)^{-1}k_2](T - r)\\ [D(I - C^2)^{-1}k_1](T - r) & [D(I - C^2)^{-1}k_2](T - r)\end{bmatrix}
$$

$$
= \operatorname{rank}\begin{bmatrix} [(I - C^2)^{-1}k_1](T - r) & [(I - C^2)^{-1}k_2](T - r) & -b_2 - [(I - C^2)^{-1}Af](T - r)\\ [D(I - C^2)^{-1}k_1](T - r) & [D(I - C^2)^{-1}k_2](T - r) & b_1 - [D(I - C^2)^{-1}Af](T - r)\end{bmatrix}
$$

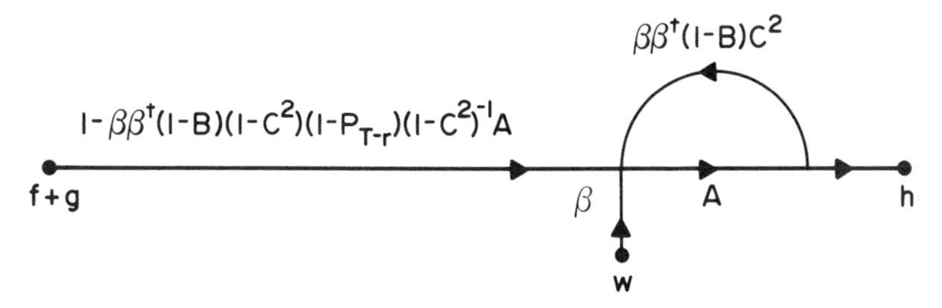

Figure 5.2

Theorem 5.1. If LOP I is regular, then the pair (u, h) in $\Omega \times G_H$ given by (5.4), (5.5) and (5.6), where $(\lambda(0), \upsilon)$ is any solution of (5.3), provides a solution.

Note that

$$u = \beta^{\dagger}(I - B)C^2 h - \beta^{\dagger}(I - B)(I - C^2)(I - P_{T-r})(I - C^2)^{-1}A(f + g)$$
$$+ \beta^{\dagger}(I - B)\{(I - (I - C^2)(I - P_{T-r})\,(I - C^2)^{-1})\}(k_1\lambda(0) + k_2\upsilon)$$

Thus the control u is a sum of three terms, a dynamic feedback, a causal (non anticipatory) function of the inputs f and g, and an open loop term w which depends on f and the constants b_1 and b_2, i.e., on some knowledge of g on the whole interval [0, T - r]. The system diagram with such a control implemented is given in Figure 5.2.

If we take r = 0, which is possible for finite dimensional state space systems, the middle term of u is missing.

Exercise. Consider the following optimization problem. For given (b_1, b_2) in $X \times X$ and (b, y_0) in $G_H \times G_H$:

Minimize on $\Omega \times G_H$: $J(u, h) = ((1/2)N_H(Au)^2 + (1/2)N_H(C(y - y_0))^2$

Subject to: $h = A(f + g + \beta u)$,

$h(t) = b(t)$ for $T - r \le t \le T$,

$[Dh](T - r) = b_1$,

$y = \gamma h$

given that $[(I - \gamma^* C^2\gamma)^{-1}Ag](T - r) = b_2$ and $[D(1 - \gamma^* C^2\gamma)^{-1}Ag](T - r) = 0$.

Use the Lagrange Multiplier techniques to show that the optimal solution and control are given by

$$h = (I - P_{T-r})b + P_{T-r}(I - \gamma^*C^2\gamma)^{-1}\{A(f + g) - \gamma^*C^2y_0 + k_1\lambda(0) + k_2\upsilon\}$$
$$u = w_1 + w_2 + w_3$$

where

$$w_1 = ß^{-1}A^{-1}\gamma^*C^2(\gamma h - y_0)$$
$$w_2 = ß^{-1}A^{-1}(I - \gamma^*C^2\gamma)(I - P_{T-r})b$$
$$-ß^{-1}A^{-1}(I - \gamma^*C^2\gamma)(I - P_{T-r})(I - \gamma^*C^2\gamma)^{-1}\{A(f + g) - \gamma^*C^2y_0\}$$
$$w_3 = ß^{-1}A^{-1}(I - \gamma^*C^2\gamma)(P_{T-r})(I - \gamma^*C^2\gamma)^{-1}\{k_1\lambda(0) + k_2\upsilon\}$$

Also $\lambda(0)$ and υ satisfy

$$b(T - r) = [(1 - \gamma^*C^2\gamma)^{-1}Af](T - r) + b_2 - [(I - \gamma^*C^2\gamma)^{-1}\gamma^*C^2y_0](T - r)$$
$$+ [(I - \gamma^*C^2\gamma)^{-1}\{k_1\lambda(0) + k_2\upsilon\}](T - r)$$

$$b_1 = [D(I - \gamma^*C^2\gamma)^{-1}Af](T - r) - [D(I - \gamma^*C^2\gamma)^{-1}\gamma^*C^2y_0](T - r)$$
$$+ [D(I - \gamma^*C^2\gamma)^{-1}\{k_1\lambda(0) + k_2\upsilon\}](T - r)$$

Here k_1 and k_2 are as defined in (5.2).

As an illustration of LOP I consider the following example.

Example 5.1. The problem is to steer the solution of the scalar equation $h'(t) = h(t - 1) + g_1(t) + v(t)$, $t \geq 0$ with initial position $h(t) = 1$ for $-1 \leq t \leq 0$ to the terminal position $h(t) = 0$ for $1 \leq t \leq 2$. We assume limited information about g_1. The system may be written as $h = f + g + Bh + u$ or $h = A(f + g + u)$ with $f(t) = 1$ for $-1 \leq t \leq 2$, $g = Cg_1$ and $u = Cv$. Here $A = (I - B)^{-1}$ with $[Bh](t) = 0$ for $-1 \leq t \leq 0$ and $[Bh](t) = \int_{[0,t]} h(\tau - 1)\, d\tau$ for $0 \leq t \leq 2$. We assume $[(I - C^2)^{-1}Ag](1) = b_2$.

A solution is obtained by setting $D = 0$ and $b_1 = 0$ in LOP I. We obtain

$$[Af](t) = \begin{cases} f(t) & -1 \le t \le 0 \\ f(t) + \int_0^t f(\tau - 1)\, d\tau & 0 \le t \le 1 \\ f(t) + \int_0^t f(\tau - 1)\, d\tau + \int_1^t (t - \tau) f(\tau - 2)\, d\tau & 1 \le t \le 2 \end{cases}$$

$$[A^{-1}f](t) = f(t) - [Bf](t) \quad -1 \le t \le 0$$

$$[C^2h](t) = \begin{cases} 0 & -1 \le t \le 0 \\ \int_0^t (t - \tau) h(\tau)\, d\tau & 0 \le t \le 2 \end{cases}$$

$$[(I - C^2)^{-1}f](t) = \begin{cases} f(t) & -1 \le t \le 0 \\ f(t) + \int_0^t \sinh(t - \tau) f(\tau)\, d\tau & 0 \le t \le 2 \end{cases}$$

The optimal trajectory and control can be determined from equations (5.3), (5.4), and (5.5). Laplace transform methods were used to determine $(I - C^2)^{-1}$.

Next we consider a local optimization problem with the standard quadratic cost functional $(1/2)\int_{[0,T]} | u' |^2 dt + (1/2)\int_{[0,T]} | y - y_0 |^2 dt$ and component signal diagram as in Figure 1. Suppose D is in $\boldsymbol{B}$, (b_1, b_2, b) is in X^3, γ is a $d \times d$ matrix and y_0 is in G_H. We want to solve the following optimal control problem.

Local Optimization Problem II (LOP II)

Minimize on $\Omega \times G_H$: $J(u, h) = (1/2)N_H(u)^2 + (1/2)N_H(C(y - y_0))^2$

Subject to: $h = f + Bh + g + \beta u$,

$h(T) = b$,

$[Dh](T) = b_1$, and

$y = \gamma h$,

where g in G_H is fixed but unknown and satisfies $[A_1 g](T) = b_2$ and $[DA_1 g](T - r) = 0$ for $A_1 = (I - B - \beta\beta^*(I - B_1)^{-1}\gamma^*\gamma C^2)^{-1}$ and $B_1 = B^* - K(\,, T)[B^*\cdot](0)$.

In this problem, we must restrict our attention to finite dimensional state space systems, i.e., $B = \alpha C$ with α a $d \times d$ matrix and $r = 0$. With this restriction we may again use the Lagrange Multiplier Theorem to characterize the optimal trajectory and control. Note that we are assuming (B, β) is controllable.

In order that (u, h) in $\Omega \times G_H$ solve this problem, there must exist multipliers (λ, μ, υ) in $G_H \times X \times X$ such that $Q_H(u', u) + Q_H(h', \gamma^*\gamma C^*Ch - \gamma^*\gamma C^*Cy_0) + Q_H(h' - Bh' - ßu', \lambda) + Q_H(h', K(, T)\mu) + Q_H(h', D^*K(, T)\upsilon) = 0$. Consequently, $u = (I - P_0)ß^*\lambda$ and $\gamma^*\gamma C^*Ch - \gamma^*C^*Cy_0 + (I - B)^*\lambda + K(, T)\mu + D^*K(, T)\upsilon = 0$. One obtains, eliminating μ from the last equation, $\lambda = \gamma^*\gamma C^2h + \gamma^*C^* Cy_0 - k\gamma^*[C^*Cy_0](0) + B^*\lambda + k\lambda(0) - k[B^*\lambda](0) + k[D^*k\upsilon](0) - D^*k\upsilon$. Thus $\lambda = (I - B_1)^{-1}\{\gamma^*\gamma C^2h - \gamma^*C^2y_0 + k\lambda(0) + k[D^*k\upsilon](0) - D^*k\upsilon\}$.

It follows that

$$\begin{aligned} h &= f + g + Bh + (I - P_0)ßß^*\lambda \\ &= f + g + Bh + ßß^*(I - B_1)^{-1}\gamma^*\gamma C^2h \\ &\quad + ßß^*(I - P_0)(I - B_1)^{-1}\{k\lambda(0) - \gamma^*C^2y_0 + k[D^*k\upsilon](0) - D^*k\upsilon\} \\ &= A_1(f + g + ßk_1\lambda(0) + ßk_2\upsilon + ßk_3) \end{aligned} \tag{5.7}$$

where $k_1 = ß^*(I - P_0)(I - B_1)^{-1}k$, $k_2 = ß^*(I - P_0)(I - B_1)^{-1}\{k[D^*k](0) - D^*k\}$, and $k_3 = -ß^*(I - B_1)^{-1}\gamma^*C^2y_0$.

The remaining constraints yield the following linear system of equations for $\lambda(0)$ and υ:

$$\begin{aligned} b &= [A_1f](T) + b_2 + [A_1ßk_1](T)\lambda(0) + [A_1ßk_2](T)\upsilon + [A_1ßk_3](T) \\ b_1 &= [DA_1f](T) + [DA_1ßk_1](T)\lambda(0) + [DA_1ßk_2](T)\upsilon + [DA_1ßk_3](T) \end{aligned} \tag{5.8}$$

Again, the rationale for the conditions $[A_1g](T) = b_2$ and $[DA_1g](T) = 0$ becomes apparent.

The optimal control is given by

$$u = ß^*(I - B_1)^{-1}\gamma^*\gamma C^2h + k_1\lambda(0) + k_2\upsilon + k_3 \tag{5.9}$$

We may argue, as for LOP I, that if $(\lambda(0), \upsilon)$ is any solution of (5.8), h is defined by (5.7), and u is defined by (5.9), then (u, h) is the unique solution of LOP II. Thus we require that LOP II be <u>regular</u>, i.e.,

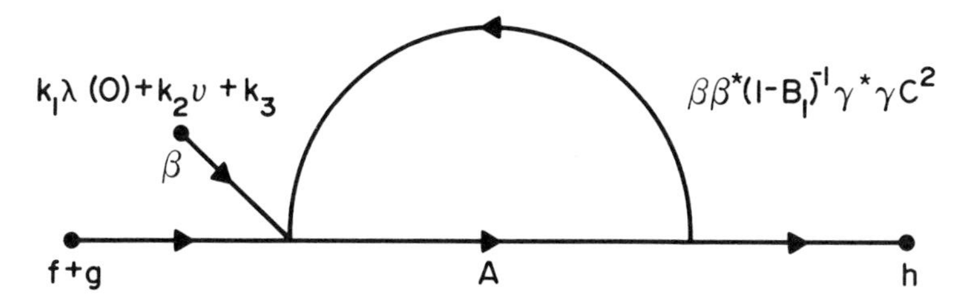

Figure 5.3

$$\text{rank}\begin{bmatrix} [A_1\beta k_1](T) & [A_1\beta k_2](T) \\ [DA_1\beta k_1(T) & [DA_1\beta k_2](T) \end{bmatrix}$$

$$= \text{rank}\begin{bmatrix} [A_1\beta k_1](T) & [A_1\beta k_2](T) & b - b_2 - [A_1 f](T) - [A_1\beta k_3](T) \\ [DA_1\beta k_1](T) & [DA_1\beta k_2](T) & b_1 - [DA_1 f](T) - [DA_1\beta k_3](T) \end{bmatrix}$$

The optimal control $u = \beta^*(I - B_1)^{-1}\gamma^*\gamma C^2 h + k_1\lambda(0) + k_2\upsilon + k_3$ is made up of two parts: a dynamic feedback and an open loop term depending on the input f, the path y_0 we want y to follow, and the constants b, b_1, and b_2. The structure of the solution is depicted in Figure 5.3.

To illustrate LOP II consider the following example.

Example 5.2. Let B = C and consider the problem of steering the scalar system $h(t) = t + [Bh](t) + 2u(t)$ for $0 \le t \le 2$ from $h(0) = 0$ to $h(2) = 1$ subject to the constraint $[Dh](T) = (1/2)\int_{[0,2]} h(t)dt = b_1$. Letting $y_0 = (1/2)I$, $b = 1$, $b_2 = 0$, $g = 0$, $y = (1/2)h$, LOP II may be used to solve this problem. One obtains using Laplace transforms,

$$[B_1 h](t) = th(0) - \int_0^t h(\tau)\, d\tau$$

$$[(I - B_1)^{-1}h](t) = h(t) - \int_0^t e^{-(t-\tau)}h(\tau)\, d\tau + (1 - e^{-t})h(0)$$

$$[A_1 h](t) = h(t) + \int_0^t \{\sqrt{2}\sinh\sqrt{2}\,(t - \tau) + \cosh\sqrt{2}\,(t - \tau)\}h(\tau)\, d\tau$$

$$[D^*K(\,,T)\upsilon](t) = [DK(\,,t)\upsilon](T) = (1 + t - t^2/4)\upsilon$$

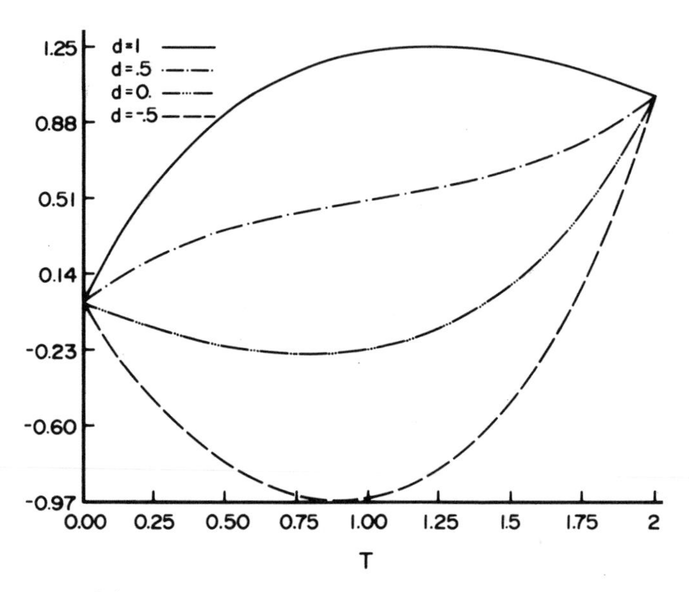

Figure 5.4

The optimal trajectory and control are determined from equations (5.7), (5.8), and (5.9). The dependence of these solutions upon the coordination parameter b_1 is depicted in Figure 5.4.

Examples 1 and 2 are indicative of the use of LOP I and LOP II. Laplace transforms can sometimes be used to determine the system operators. In more general problems, our Hilbert space setting aids in the representation and approximation of the operators, see Chapter III.

3. A Coordination Strategy

We will present a control coordination strategy, a scheme for coordinating local control decisions, for large scale systems in which component interactions are suitably limited. First, consider the two component control system

$$h_1 = f_1 + B_1h_1 + C_2h_2 + \beta_1u_1$$
$$h_2 = f_2 + C_1h_1 + B_2h_2 + \beta_2u_2$$

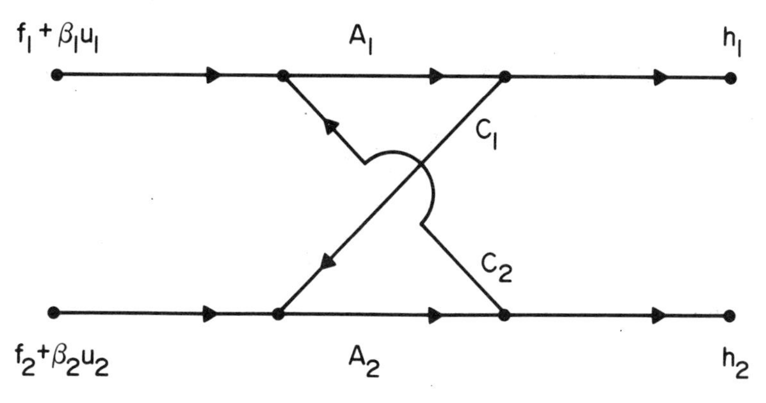

Figure 5.5

where the B's and C's are from $\boldsymbol{B}$. Assume both (B_1, β_1) and (B_2, β_2) are controllable. We can diagram the system as seen in Figure 5.5.
with $A_i = (I - B_i)^{-1}$ for $i = 1, 2$.

We say the two components are stably connected provided for all g in P_0G_H the condition $C_i(I - C^2)^{-1}A_iC_jg = 0$ holds for $\{i, j\} = \{1, 2\}$. Large scale systems composed of several components, each pair of which is stably connected, will be referred to as stably connected systems [12]. Local control laws for stably connected systems can be coordinated by imposing side conditions as in the previous section and exchanging that information among the component controllers.

The stably connected condition can be given a geometric interpretation in specific cases. If $C_i = 0$ for $i = 1$ or 2 then the components are stably connected and there is a hierarchical relation between the components. If C_i commutes with $(I - C^2)^{-1}$ for $i = 1$ and 2 on P_0G_H then the stably connected condition is equivalent to the condition $C_iA_iC_j = 0$ for $\{i, j\} = \{1, 2\}$. If the latter condition is not satisfied, one could argue that the input-output operator for the first component should be $(I - A_1C_2A_1C_1)^{-1}A_1$ rather than A_1. The diagram for the first component would then have the form in Figure 5.6.

Exercise. For $[C_ih](t) = \alpha\int_{[0,t]} h(\tau)\, d\tau$ or $[C_i](t) = \alpha\int_{[0,t]} h(\tau\text{-}r)\, d\tau$ for $0 \leq t$ where α is a $d \times d$ matrix verify that C_i commutes with $(I - C^2)^{-1}$ on P_0G_H.

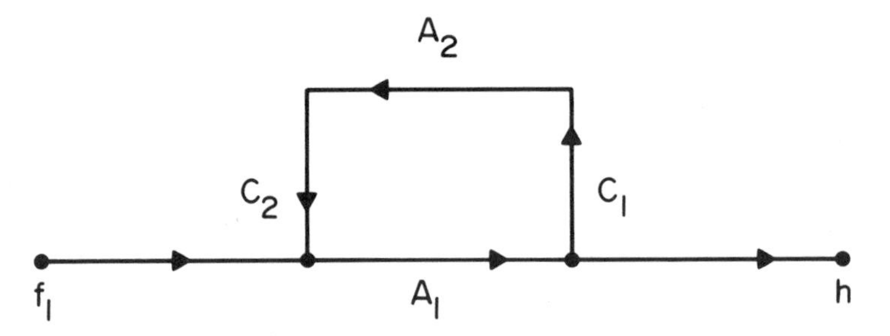

Figure 5.6

For finite dimensional state space systems i.e., $r = 0$ and $B_1 = \alpha_{11}C$, $C_2 = \alpha_{12}C$, $C_1 = \alpha_{21}C$, and $B_2 = \alpha_{22}C$ with the α_{ij} matrices of appropriate dimensions the two components will be stably connected provided controllability subspace [13] determined by $\{\alpha_{ii}, \alpha_{ij}\}$ is a subspace of the null space of α_{ji} for $\{i, j\} = \{1, 2\}$. Further comments related to the stably connected condition for finite dimensional systems will be given at the end of this section.

For a stably connected two component system as described in Figure 5.5, consider the problem of local controllers choosing controls u_1 and u_2, respectively, on the interval $0 \leq t \leq T$ which steer h_1 and h_2 to zero over the interval $T - r \leq t \leq T$, i.e., LOP I. In order that each control can be determined without full knowledge of other component control decision, we are led to the following control coordination strategy.

Coordination Strategy. Given elements c_1 and c_2 from X and f_1 and f_2 in G_H the first controller solves LOP I with $A = A_1$, $D = D_1 = (I - C^2)^{-1}A_2C_1$, $b_1 = c_1$ and $b_2 = c_2$. The second controller solves LOP I with $A = A_2$, $D = D_2 = (I - C^2)^{-1}A_1C_2$, $b_1 = c_2$, and $b_2 = c_1$.

In this strategy the condition $[(I - C^2)^{-1}A_1C_2h_2\}(T - r) = c_2$ represents the information that the first controller must have about the influence of the second component, i.e., C_2h_2, in order to solve its local optimization problem. This condition becomes a constraint on the second component. Similarly, the condition $c_1 = \{(I - C^2)^{-1}A_2C_1h_1\}(T - r)$ becomes a constraint on the first component. The stably

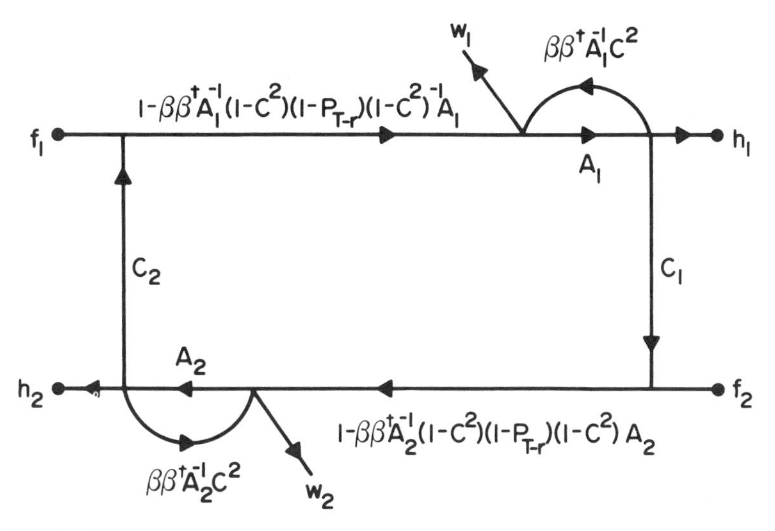

Figure 5.7

connected assumption implies that $[D_i(I - C^2)^{-1}A_iC_jh_j](T - r) = [D_iD_jh_j](T - r) = 0$ for each component and each controller can solve its local optimization problem provided the problem is regular. The controlled system becomes Figure 5.7 where the w_i's depend on the f_i's and b's as before. An implementation of the strategy consists of the w's (or better the c's) and the "hardwired" operators.

The implementation of our coordination strategies for a two component system requires that appropriate values for the parameters c_1 and c_2 be available. These parameters are referred to as coordination parameters and the problem of determining values for these parameters leads to a higher level coordination problem. We will return to this problem in Section 4.

We turn next to the problem of applying these results to finite dimensional state space systems. The basic idea is to decompose a given system into components which are stably connected.

A Basic Decomposition. We now present a basic decomposition procedure [11] for finite dimensional systems of the form $h(t) = f(t) + \alpha\int_{[0,t]} h(s)\,ds + \beta_1u_1 + \beta_2u_2$ where $r = 0$, α is a $d \times d$ matrix and β_i is a $d \times n_i$ matrix for $i = 1, 2$. For $i =$

1, 2 let Π_i denote the projection of X onto the controllability subspace spanned by the columns of the matrix $\{\beta_i, \alpha\beta_i, \ldots, \alpha^{d-1}\beta_i\}$ and Π_3 a projection of X onto $\Pi_1 X \cap \Pi_2 X$. Assume $X = \Pi_1 X + \Pi_2 X$ and recall that the identity may be written as $I = \Pi_1 + \Pi_2 - \Pi_3$.

We want a decomposition of the system in terms of the Π's to which we can apply our distributed control strategy. For $0 \leq m_0,\ m_1,\ m_2 \leq 1$, let

$$f_1 = (\Pi_1 - m_0\Pi_3)f$$
$$f_2 = (\Pi_2 - (1 - m_0)\Pi_3)f$$

and

$$\begin{pmatrix} \alpha_{11} & \alpha_{12} \\ \alpha_{21} & \alpha_{22} \end{pmatrix} = \begin{pmatrix} (\Pi_1 - m_1\Pi_3)\alpha\Pi_1 & m_2\Pi_3\alpha\Pi_2 \\ m_1\Pi_3\alpha\Pi_1 & (\Pi_2 - m_2\Pi_3)\alpha\Pi_2 \end{pmatrix}$$

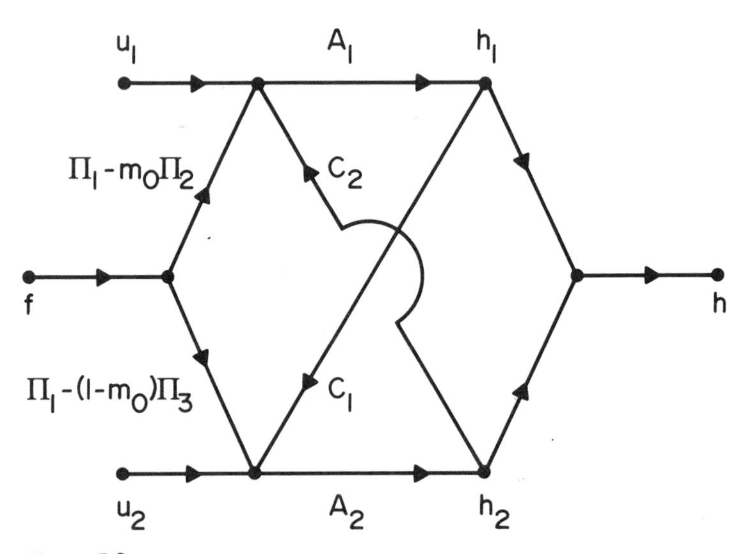

Figure 5.8

Furthermore, let $B_1 = \alpha_{11}C$, $C_1 = \alpha_{21}C$, $B_2 = \alpha_{22}C$, $C_2 = \alpha_{12}C$ and $K(u, v) = \{1 + \min(u, v)\}I_d$ for $0 \le v$, u. Our decomposition has the form

$$h_1 = f_1 + B_1h_1 + C_2h_2 + \beta_1u_1$$
$$h_2 = f_2 + C_1h_1 + B_2h_2 + \beta_2u_2$$

It follows using the invariance properties of the controllability subspaces that

$$h_1 + h_2 = f_1 + f_2 + (\alpha_{11} + \alpha_{21})\int_0^t h_1\, dt + (\alpha_{12} + \alpha_{22})\int_0^t h_2\, dt + \beta_1u_1 + \beta_2u_2$$

$$= f + \Pi_1\alpha\Pi_1\int_0^t h_1\, dt + \Pi_2\alpha\Pi_2\int_0^t h_2\, dt + \beta_1u_1 + \beta_2u_2$$

$$= f + \Pi_1\alpha\int_0^t h_1\, dt + \Pi_2\alpha\int_0^t h_2\, dt + \beta_1u_1 + \beta_2u_2$$

$$= f + (\Pi_1 - \Pi_3)\alpha\int_0^t h_1\, dt + \Pi_3\alpha\int_0^t h_1\, dt$$

$$+ (\Pi_2 - \Pi_3)\alpha\int_0^t h_2\, dt + \Pi_3\alpha\int_0^t h_2\, dt + \beta_1u_1 + \beta_2u_2$$

$$= f + (\Pi_1 + \Pi_2 - \Pi_3)\alpha\int_0^t (h_1 + h_2)\, dt + \beta_1u_1 + \beta_2u_2$$

Thus $h = h_1 + h_2$ and the decomposition can be diagramed as in Figure 5.8 where $A_i = (I - B_i)^{-1}$, $i = 1, 2$.

The system will be stably connected provided $C_1A_1C_2 = C_2A_2C_1 = 0$. Here $C_2A_2C_1 = m_1\Pi_3\alpha\Pi_1(I - (\Pi_2 - m_2\Pi_3)\alpha\Pi_2C)^{-1}m_2\Pi_3\alpha\Pi_2$. A similar expression holds for $C_1A_1C_2$.

We consider three cases i) $\Pi_1X \cap \Pi_2X = \{0\}$; ii) $\Pi_2X \subseteq \Pi_1X$; and iii) $\Pi_1X \neq X$, $\Pi_2 X \neq X$, and $\Pi_1X \cap \Pi_2X \neq \{0\}$. If $y = \{I - (\Pi_2 - m_2\Pi_3)\alpha\Pi_2C\}^{-1} m_2\Pi_3\alpha\Pi_2x$ then the invariance properties of the controllability subspaces imply that $\Pi_1Cy = \Pi_3Cy$. Consequently, the stably connected condition is satisfied for all choices of m_1 and m_2 whenever $\Pi_3\alpha\,\Pi_3 = 0$. In the first case $\Pi_3 = 0$ and there is no interaction between the components. In the second case $\Pi_3 = \Pi_2$ and the situaion seems to be most interesting when $\Pi_3\alpha\Pi_3 \neq 0$. In this case the stably connected condition can be met by choosing either m_1 or m_2 to be zero

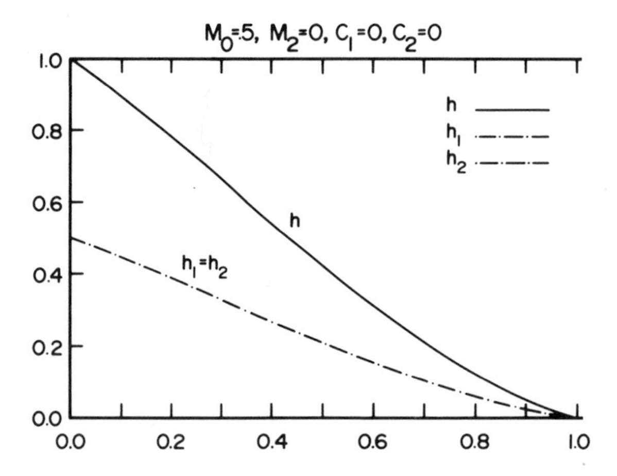

Figure 5.9

forming a "hierarchical" system. Finally, the third case is mixed since $\Pi_3\alpha\Pi_3$ might or might not be zero. When $\Pi_3\alpha\Pi_3 = 0$ and $m_1 m_2 \neq 0$ note that the system is not a hierarchy and the components interact.

The following two examples illustrate the rich class of systems exposed by the basic decomposition.

Example 5.3 [11]. Let $S = [0, 1]$, $d = 1$, $\alpha = 1$, and $\beta_1 = \beta_2 = 1$. Then $\Pi_1 = \Pi_2 = \Pi_3 = 1$. Here the original system is assumed to have the form $h = f + \alpha\int_{[0,t]} h(s)ds + u_1 + u_2$. If $f = 1$ then $f_1 = 1 - m_0$ and $f_2 = m_0$. Since $\Pi_3\alpha\Pi_3 = 1 \neq 0$, we choose $m_1 = 0$. Hence

$$\alpha_{11} = 1 \qquad \alpha_{12} = m_2$$
$$\alpha_{21} = 0 \qquad \alpha_{22} = 1 - m_2$$

Each component controller solves a version of LOP I. The operators involved in the explicit formulas for the responses h_1 and h_2 (see Theorem 5.1) namely, for $0 \le t \le 1$ and (f_1, f_2) in $G_H \times G_H$,

$$h_1(t) = [(I - C^2)^{-1}\{A_1(f_1 + C_2 h_2) + \lambda_1 I\}](t)$$
$$h_2(t) = [(I - C^2)^{-1}\{A_2 f_2 + \lambda_2 I + [D_2{}^*K(\ ,1)\upsilon](0)I - (I - P_0)D_2{}^*K(\ ,1)\upsilon\}](t)$$

have straightforward representations. For f in G_H and t in S,

$$[A_1 f](t) = f(t) + \int_0^t e^{t-s} f(s)\, ds$$

$$[A_2 f](t) = f(t) + (1 - m_2) \int_0^t \exp((1 - m_2)(t - s)) f(s)\, ds,$$

$$[(1 - C^2)^{-1} f](t) = f(t) + \int_0^t \sinh(t - s) f(s)\, ds,$$

$$D_1 = 0$$

$$[C_2 f](t) = m_2 \int_0^t f(s)\, ds$$

$$D_2 = (I - C^2)^{-1} A_1 C_2$$

For each t in S,

$$[D_2{}^* K(\ ,1)](t) = [D_2 K(\ ,t)](1)$$
$$= m_2 \int_0^t \cosh(1 - s)(2e^s - 1)\, ds + m_2 \int_t^1 \cosh(1 - s)(2e^s - e^{s-t})\, ds$$

Using these operator representations, we can calculate component responses and controls. If $m_0 = 1/2$ and $m_1 = m_2 = 0$ there is no interaction between the com-

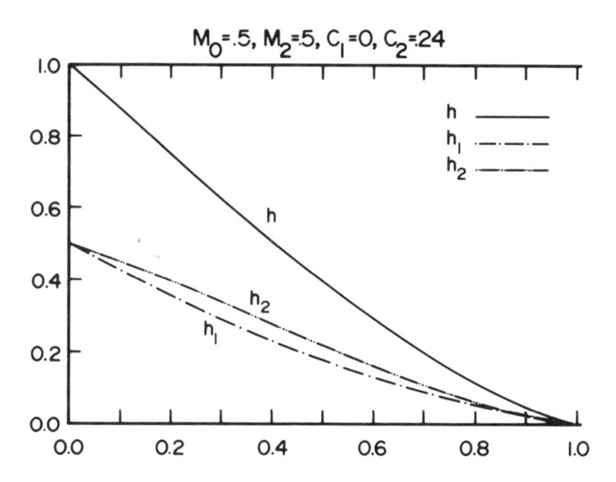

Figure 5.10

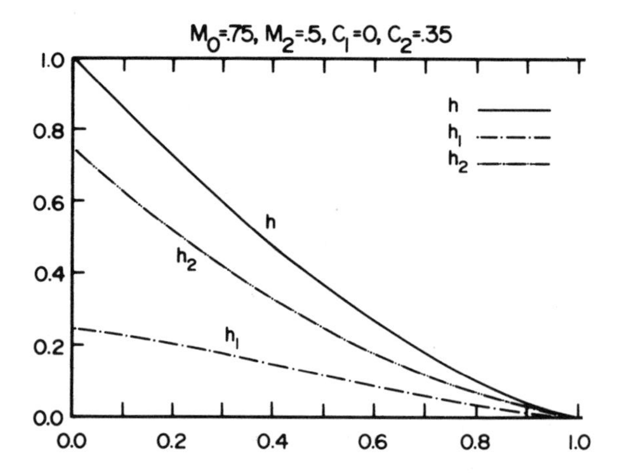

Figure 5.11

ponents and $h_1 = h_2$. Also, there is no difference in the response of the system under decentralized control and the system under centralized control, see Figure 5.9.

However, in Figure 5.10 with $m_0 = 1/2$, $m_1 = 0$, and $m_2 = 1/2$ the c's which appear in the added constraints of the control strategy have to be chosen judiciously to achieve a system response comparable to the response of the system under centralized control. This same point for $m_0 = 3/4$, $m_1 = 0$, and $m_2 = 1/2$ is illustrated in Figure 5.11.

Example 5.4 [11]. For a second example where $\Pi_3 \neq 0$ and $\Pi_3 \alpha \Pi_3 = 0$, let

$$d = 3,\ \alpha = \begin{bmatrix} 1 & 0 & 0 \\ 2 & 0 & -3 \\ 0 & 0 & 4 \end{bmatrix},\ \beta_1 = \begin{bmatrix} 1 & 0 \\ 0 & 1 \\ 0 & 0 \end{bmatrix},\ \text{and } \beta_2 = \begin{bmatrix} 0 & 0 \\ 1 & 0 \\ 0 & 1 \end{bmatrix}$$

We can choose

$$\Pi_1 = \begin{bmatrix} 1 & 0 & 0 \\ 0 & 1 & 0 \\ 0 & 0 & 0 \end{bmatrix},\ \Pi_2 = \begin{bmatrix} 0 & 0 & 0 \\ 0 & 1 & 0 \\ 0 & 0 & 1 \end{bmatrix},\ \text{and } \Pi_3 = \begin{bmatrix} 0 & 0 & 0 \\ 0 & 1 & 0 \\ 0 & 0 & 0 \end{bmatrix}$$

Note that $\Pi_3 \alpha \Pi_3 = 0$.

$$\alpha_{11} = \begin{bmatrix} 1 & 0 & 0 \\ 2(1-m_1) & 0 & 0 \\ 0 & 0 & 0 \end{bmatrix} \qquad \alpha_{12} = \begin{bmatrix} 0 & 0 & 0 \\ 0 & 0 & -3m_2 \\ 0 & 0 & 0 \end{bmatrix}$$

$$\alpha_{21} = \begin{bmatrix} 0 & 0 & 0 \\ 2m_1 & 0 & 0 \\ 0 & 0 & 0 \end{bmatrix} \qquad \alpha_{22} = \begin{bmatrix} 0 & 0 & 0 \\ 0 & 0 & -3(1-m_2) \\ 0 & 0 & 4 \end{bmatrix}$$

$$\text{If } f = \begin{bmatrix} 1 \\ 1 \\ 1 \end{bmatrix} \text{ then } f_1 = \begin{bmatrix} 1 \\ 1-m_0 \\ 0 \end{bmatrix} \text{ and } f_2 = \begin{bmatrix} 0 \\ m_0 \\ 1 \end{bmatrix}$$

As in Example 3, local controllers can solve a form of LOP I in order to steer $h(0) = f(0)$ to $h(1) = 0$.

Next, we look to the representations of A_i, C_i, and $D_i{}^*K(\ ,1)y$, for $i = 1, 2$ and y in X which arise in LOP I. For g in G_H and t in S

$$[A_1g](t) = \begin{cases} g_1(t) + \int_0^t e^{t-s} g_1(s)\,ds \\ g_2(t) + 2(1-m_1)\int_0^t e^{t-s} g_1(s)\,ds \\ g_3(t) \end{cases}$$

$$[A_2g](t) = \begin{cases} g_1(t) \\ g_2(t) - 3(1-m_2)\int_0^t e^{4(t-s)} g_3(s)\,ds \\ g_3(t) + 4\int_0^t e^{4(t-s)} g_3(s)\,ds \end{cases}$$

$$[C_1g](t) = \begin{cases} 0 \\ 2m_1\int_0^t g_1(s)\,ds \\ 0 \end{cases}, \qquad \text{and} \qquad [C_2g](t) = \begin{cases} 0 \\ -3m_2\int_0^t g_3(s)\,ds \\ 0 \end{cases}$$

If t is in S and x is in X then

$$\langle [D_1{}^*K(\ ,1)y](t), x\rangle = 2m_1\{1 + t - t^2/2 + \int_0^t \sinh(1-s)(s+s^2)\,ds$$
$$+ \int_t^1 \sinh(1-s)(s + st - t^2/2)\,ds\}y_2x_1$$

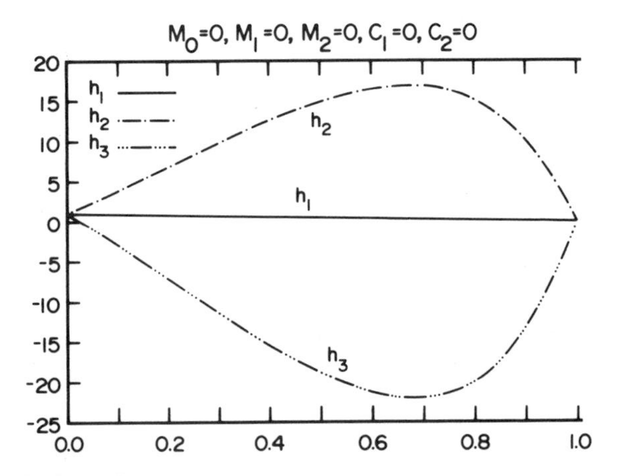

Figure 5.12

and

$$<[D^2*K(\ ,1)y](t), x> = -3m_2\{1 + t - t^2/2 + \int_0^t \sinh(1 - s)(s + s^2/2)\,ds$$

$$+ \int_t^1 \sinh(1 - s)(s + st - t^2/2)\,ds\}y_2x_3$$

In Figure 5.12 with $m_0 = m_1 = m_2 = 0$, i.e., with no interaction among the components, we see that local optimization leads to values for h_2 and h_3 of large magnitude. However, in Figure 5.13 with $m_0 = m_1 = 0$ and $m_2 = .5$ the values of h_2 and h_3 are contained in a much smaller interval.

An hereditary system example. Consider the two tank system sketched in Figure 5.14.

The sketch might represent either a real system or an abstraction; for instance, rooms of the house discussed earlier. Solution is pumped into the system at a and e, there are interconnections at c and d, and solution is pumped out at b and f. Assume the flow rates v_a through v_f are known constants with $v_a + v_d = v_c + v_b$ and $v_c + v_e = v_d + v_f$. The concentration of the solute at a is a function $x_1 + w_1$ of time, where x_1 is a known input and w_1 is to be chosen by the first controller. Similarly for x_2 and w_2 at e.

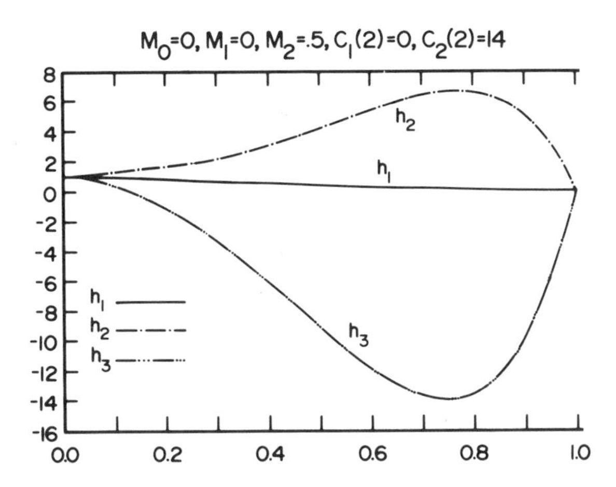

Figure 5.13

Let h_i be the total amount of the solute in tank T_i with volume V_i. If the solution is assumed homogeneous, well mixed, then the concentration is simply h_i/V_i. However, we want to substitute for this the assumption that the concentrations at b and c are given at time t by $(1/V_1)\int_{[-r,0]} \phi_b(\tau)h_1(\tau + t)\, d\tau$ and $(1/V_1) \cdot \int_{[-r,0]} \phi_c(\tau)h_1(\tau + t)\, du$, respectively, where $r > 0$, ϕ_b and ϕ_c are given nonnegative functions, and $\int_{[-r,0]} \phi_b(\tau)\, dt = \int_{[-r,0]}\phi_c(\tau)\, dt = 1$. Similarly for d and f. In this model of "mixing" the concentration of the solute at an outlet at time $t \geq 0$ is taken to be a weighted average of h_i/V_i over the interval $[t - r, t]$.

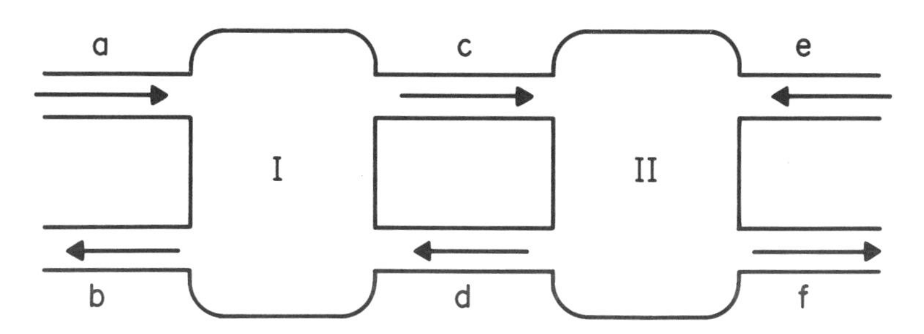

Figure 5.14

Our model becomes the hereditary system

$$dh_1/dt = v_a(x_1(t) + w_1(t)) - (v_b/V_1)\int_{-r}^{0}\phi_b(\tau)h_1(\tau + t)\,d\tau$$

$$- (v_c/V_1)\int_{-r}^{0}\phi_c(\tau)\,h_1(\tau + t)\,d\tau + (v_d/V_2)\int_{-r}^{0}\phi_d(\tau)h_2(\tau + t)\,d\tau$$

$$dh_2/dt = v_e(x_2(t) + w_2(t)) + (v_c/V_1)\int_{-r}^{0}\phi_c(\tau)h_1(\tau + t)\,d\tau$$

$$- (v_d/V_2)\int_{-r}^{0}\phi_d(\tau)\,h_2(\tau + t)\,d\tau - (v_f/V_2)\int_{-r}^{0}\phi_f(\tau)h_2(\tau + t)\,d\tau$$

We want to start controlling the system at time $t = 0$; but it is necessary to specify x_1, x_2, h_1, and h_2 on $[-r, 0]$. We will assume that x_1 and x_2 are constant and $x_1(t) = x_2(t) = h_1(t)/V_1 = h_2(t)/V_2 = m_0$ for $-r \le t \le 0$, i.e., that the system is in equilibrium. The objective is to achieve a new equilibrium m_T at time $t = T$. In order to achieve this we must have $h_1(t)/V_1 = h_2(t)/V_2 = m_T$ for $T - r \le t \le T$.

The ith controller is to choose w_i to force h_i to its objective $m_T V_i$ on $[T - r, T]$. The nature of the coupling between the components suggests that this cannot be done without taking into consideration the actions of the other controller. Furthermore, for the ith controller to optimize he must know something about the response of the other system on the whole interval $[-r, T]$ at $t = 0$.

Consider an integral of the form $\int_{[-r,0]}\phi(u)f(t + u)\,du$. Note that

$$\int_{-r}^{0}\phi(u)f(t + u)\,du = \int_{t-r}^{t}\phi(-(t - u))f(u)\,du$$

$$= \begin{cases} \int_0^t g(t - u)f(u)\,du & r \le t \le T \\ \int_0^t g(t - u)f(u)\,du + \int_{t-r}^{0}\phi(-(t - u))f(u)\,du & 0 \le t \le r \end{cases}$$

where

$$g(u) = \begin{cases} \phi(-u) & 0 \le u \le r \\ 0 & \text{otherwise} \end{cases}$$

Thus our model for the two tank flow systems can be written abstractly as

$$h_1 = f_1 + u_1 + B_1h_1 + C_2h_2$$
$$h_2 = f_2 + u_2 + C_1h_1 + B_2h_2$$

where f_i, u_i, B_i, and C_i, $i = 1, 2$, are defined as follows. For $-r \le t \le T$ let

$$f_1(t) = \begin{cases} m_0V_1 & -r \le t \le 0 \\ m_0\{V_1 + v_at + \int_0^t\int_{s-r}^0 - v_b\phi_b(-(s-\tau)) - v_c\phi_c(\tau - s) \\ \qquad + v_d\phi_d(\tau - s)\, d\tau\, ds\} & 0 \le t \le r \\ m_0\{V_1 + v_at + \int_0^r\int_{s-r}^0 - v_b\phi_b(\tau - s) - v_c\phi_c(\tau - s) \\ \qquad + v_d\phi_d(\tau - s)\, d\tau\, ds\} & r \le t \le T \end{cases}$$

$$[B_1h](t) = \begin{cases} 0 & -r \le t \le 0 \\ \int_0^t g_{11}(t-s)h(s)\, ds & 0 \le t \le T \end{cases}$$

where

$$g_{11}(t) = \begin{cases} \int_0^t - (v_b/V_1)\phi_b(-s) - (v_c/V_1)\phi_c(-s)\, ds & 0 \le t \le r \\ g_{11}(r) & r \le t \le T \end{cases}$$

$$[C_2h](t) = \begin{cases} 0 & -r \le t \le 0 \\ \int_0^t g_{12}(t-s)\, h(s)\, ds & 0 \le t \le T \end{cases}$$

where

$$g_{12}(t) = \begin{cases} \int_0^t (v_d/V_2)\phi_d(-r)\, ds & 0 \le t \le r \\ g_{12}(r) & r \le t \le T \end{cases}$$

and

$$u_1(t) = \begin{cases} 0 & -r \le t \le 0 \\ v_a\int_0^t w_1(r)\, ds & 0 \le t \le T \end{cases}$$

Note that for g and h in G and t in [0, T]

$$\int_0^t \int_0^s g(s - \tau)\, h(\tau)\, d\tau\, ds = [(g * h) * 1](t)$$
$$= [(g*h)*1](t)$$
$$= [(g*1)*h](t)$$
$$= [G*h](t)$$

where * denotes ordinary convolution and

$$G(t) = [Cg](t)$$

The functions f_2 and u_2 and the operators C_1 and B_2 are defined in a similar manner.

Exercise. For i = 1, 2, $C_i(I - C^2)^{-1} = (I - C^2)^{-1}C_i$, i.e., the stably connected condition reduces to $C_1A_1C_2 = C_2A_2C_1 = 0$.

A numerical example. The results of a numerical example are presented graphically. For the example, choose in the first formulation of the tank system $V_1 = 1$, $V_2 = 2$, $v_a = v_c = v_f = 2$, $v_b = v_d = v_e = 1$, $m_0 = 1$, and $m_T = 0$. Furthermore, let

$$\phi_b(-u) = \phi_c(-u) = \begin{cases} 144(u - 1/3) & 1/3 \le u \le 5/12 \\ -144(u - 5/12) + 12 & 5/12 \le u \le 1/2 \\ 0 & \text{otherwise} \end{cases}$$

and

$$\phi_d(-u) = \phi_f(-u) = \begin{cases} 36(u - 2/3) & 2/3 \le u \le 5/6 \\ -36(u - 5/6) + 6 & 5/6 \le u \le 1 \\ 0 & \text{otherwise} \end{cases}$$

Exercise. Show that the system is stably connected provided r = 1 and T = 2.

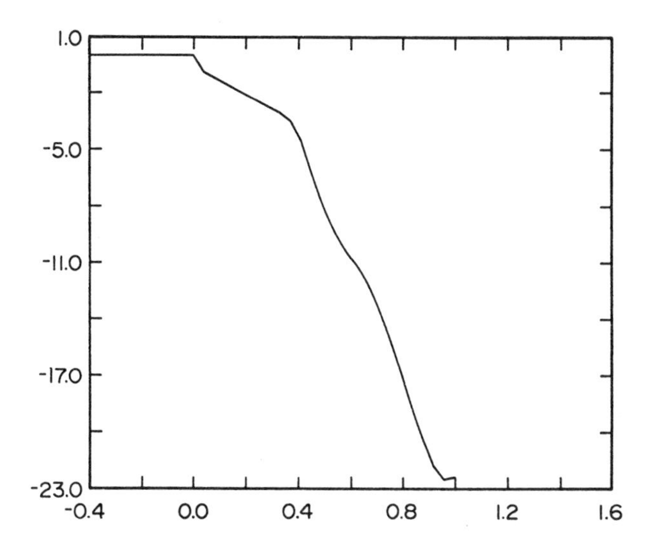

FIRST CONTROL WHEN SECOND CONTROLLER DOES NOTHING

Figure 5.15

In Figure 5.15 we see that the controller of the smaller tank T_1 must exert a large effort to move his component from $h_1(0) = 1$ to $h_1(1) = 0$ if the second controller does nothing. On the other hand, Figure 5.16 shows in this case that tank T_2 moves from $h_2(0) = 2$ to $h_2(1) \approx -14.2$. Thus we might expect that the components would benefit by cooperating. In Figures 5.17-5.19 we see the effects of various choices for the coordination parameters. First, note that the system responses are sensitive to the choices. Second, in the absence of a system cost/benefit function we cannot say which choice is preferred.

More General Networks. Coordinating strategies can be developed for any stably connected system. As previously mentioned, physical limitations on component interactions frequently result in stably connected systems. A rich class of such systems are those which can be thought of as hierarchies. A canonical example is a multiple component system with signal diagram in the form of Figure 5.20. Clearly each pair of components is stably connected. We can envision a command structure

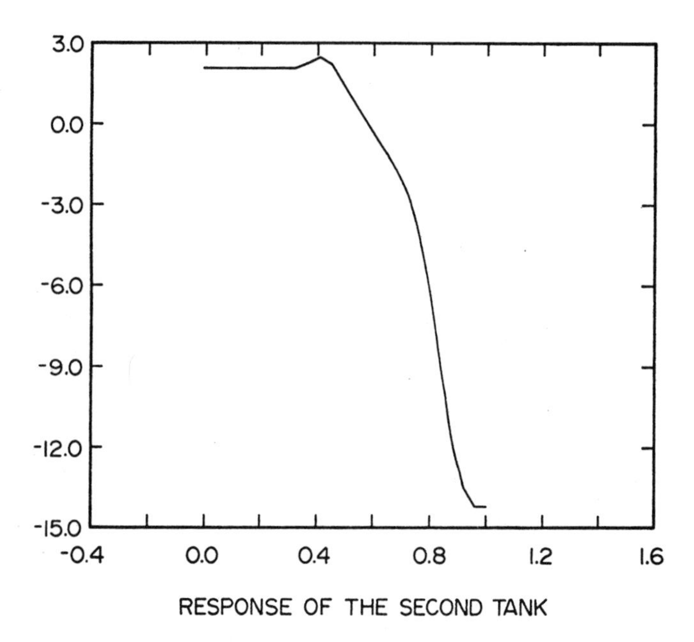

Figure 5.16

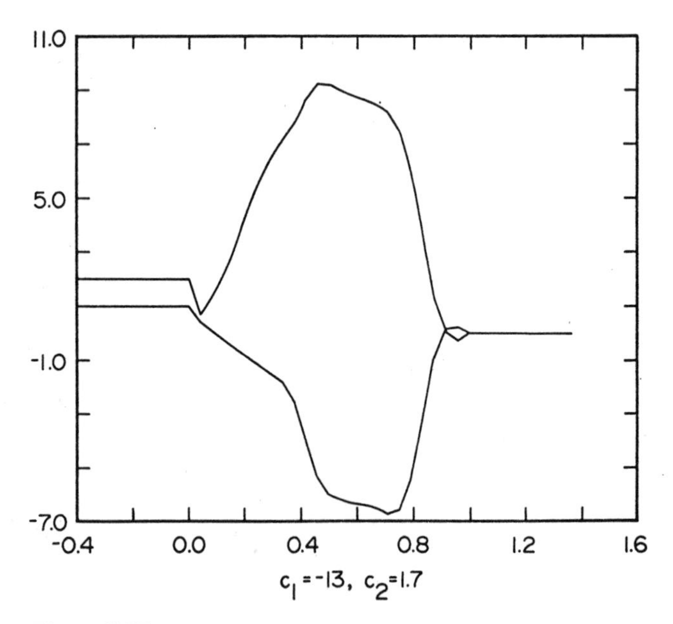

Figure 5.17

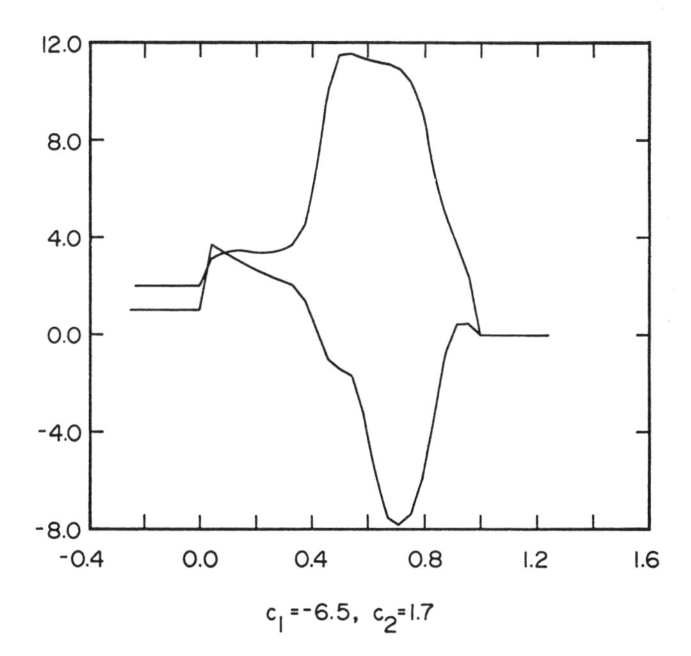

Figure 5.18

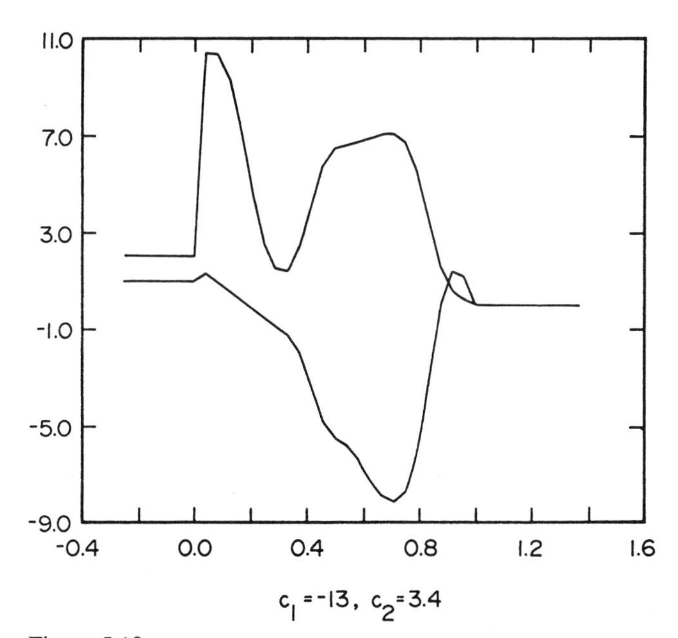

Figure 5.19

where the controller for the first component sets the coordination parameters and hence the side conditions for the other two components, i.e., specifies the values of $[(I-C^2)^{-1}A_1C_2h_2](T)$ and $[(I-C^2)^{-1}A_1C_3h_3](T)$ if LOP I is used. More generally, we can consider a grid network in the form of Figure 5.21 provided the system is stably connected. The side conditions, using LOP I, would have the following form.

On the first component: $[(I-C^2)^{-1}A_2C_1h_1](T) = c_1$
On the second component: $[(I-C^2)^{-1}A_1C_{21}h_2](T) = c_{21}$ and
$[(I-C^2)^{-1}A_3C_{23}h_2](T) = c_{23}$
On the third component: $[(I-C^2)^{-1}A_2C_3h_3](T) = c_3$

4. System Coordination

A control coordination strategy for decentralized control is a scheme which allows some level of autonomous component control. For systems whose component interactions are suitably limited, we have shown that a variety of coordination strategies are possible. Our approach is to add conditions to the component requirements and to arrange exhanges of information which enable all of the components to achieve their objectives independently. It should be noted that component interactions are taken into consideration at this stage and the desired terminal conditions are achieved in one step by the solutions of the separate optimization problems.

In order to implement such control strategies, one may envision a coordinator who acts or may act upon the system by setting the component objectives, terminal

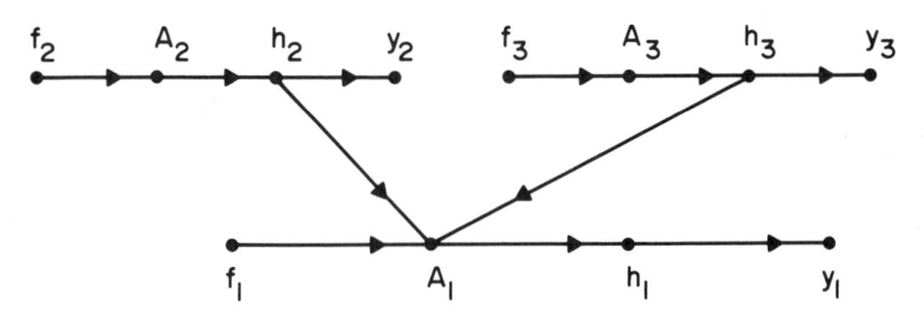

Figure 5.20

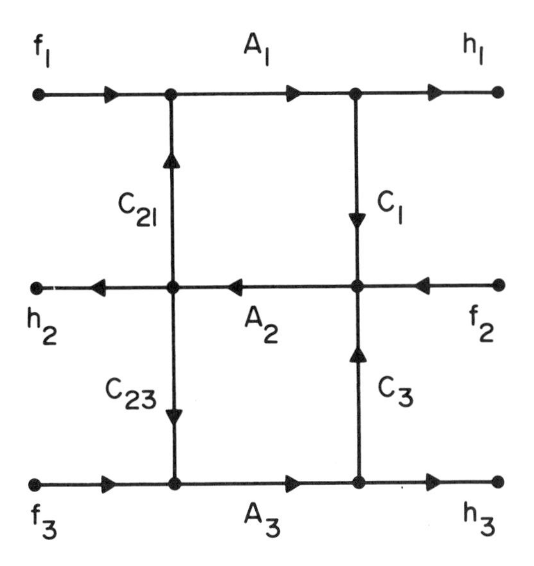

Figure 5.21

constraints, desired trajectories, or side conditions necessary for coordination. Such a coordinator may use centralized knowledge of each component to determine a coordination strategy. Realistically, only limited information on component performance would be available to the coordinator.

Alternately, coordination can be achieved by the imposition of a command structure among the system components for the choice of component objectives and the passage of information.

In our coordination strategies, once additional component constraints have been delineated, there remains the problem of determining appropriate values for the coordination parameters which have been introduced. Again, there are many possibilities. A coordinator could use centralized knowledge of the dependence of component performance or responses in order to determine appropriate values for these parameters. Another possibility is to allow certain components to treat these parameters as design parameters while the remaining components would have extra side conditions placed upon their requirements.

To illustrate some of the possibilities for this higher level coordination problem we take up the following example.

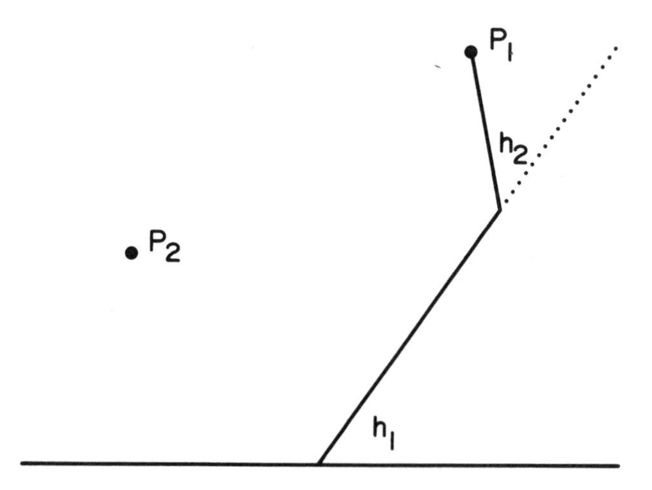

Figure 5.22

Consider the control of a two link mechanical arm. It is assumed that the motion takes place in the x-y plane and that two motors are used to control the link angles h_1 and h_2, see Figure 5.22. Counter clockwise rotation corresponds to the positive direction for angle measurement. The system objective is to move the end of link two from point P_1 to point P_2. Consequently, the angles h_1 and h_2 must vary in a coordinated manner.

Assuming first order models for each controller, the problem is to steer the components

$$dh_1/dt = b_1 - h_1 - h_2 + v_1$$
$$dh_2/dt = b_2 - h_2 + v_2$$

with observation processes $y_1 = h_1$ and $y_2 = h_2$ from initial angles $h_1(0) = a_1$, $h_2(0) = a_2$ to terminal angles $h_1(T) = b_1$, $h_2(T) = b_2$.

An integration permits us to rewrite the system in the form

$$h_1 = f_1 + B_1h_1 + C_2h_2 + u_1$$
$$h_2 = f_2 + B_2h_2 + u_2$$

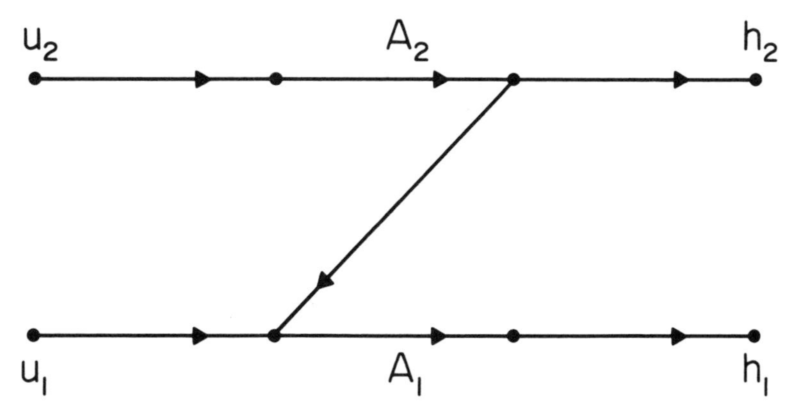

Figure 5.23

where $f_i(t) = a_i + b_i t$, $u_i(t) = Cv_i$, $B_i h = -Ch$ for $i = 1, 2$, and $C_2 h = -Ch$ with $[Ch](t) = \int_{[0,t]} h(\tau)\, d\tau$ for $0 \le t \le T$. Let $S = [0, T]$ and $K(u, v) = 1 + u$ for $0 \le u \le v \le T$ and $K(u, v) = 1 + v$ for $0 \le v \le u \le T$. The signal diagram for the system is pictured in Figure 5.23 with $A_1 = [I - B_1]^{-1}$ and $A_2 = [I - B_2]^{-1}$.

It is assumed that each controller knows the local system structure and is to use this knowledge along with the given observations as a basis for the choice of a control law. Our coordination strategies suggest the following problem for each controller, see LOP II in Section 2.

I. Minimize on $\Omega \times G_H$: $J(u_1, h_1) = (1/2)N_H(C(h_1 - b_1))^2 + (1/2)N_H(u_1)^2$
Subject to: $h_1 = f_1 + B_1 h_1 + u_1$
$h_1(T) = b_1$

given $[A_{11} C_2 h_2](T) = c$ where $A_{11} = (I - B_1 - ((I - B_{11})^{-1} C^2)^{-1}$ and $B_{11} = B_1{}^* - K(\ ,T)[B_1{}^* \bullet](0)$.

II. Minimize on $\Omega \times G_H$: $J(u_2, h_2) = (1/2)N_H(C(h_2 - b_2))^2 + (1/2)N_H(u_2)^2$
Subject to: $h_2 = f_2 + B_2 h_2 + u_2$
$[A_{11} C_2 h_2](T) = c$

with A_{11} as above.

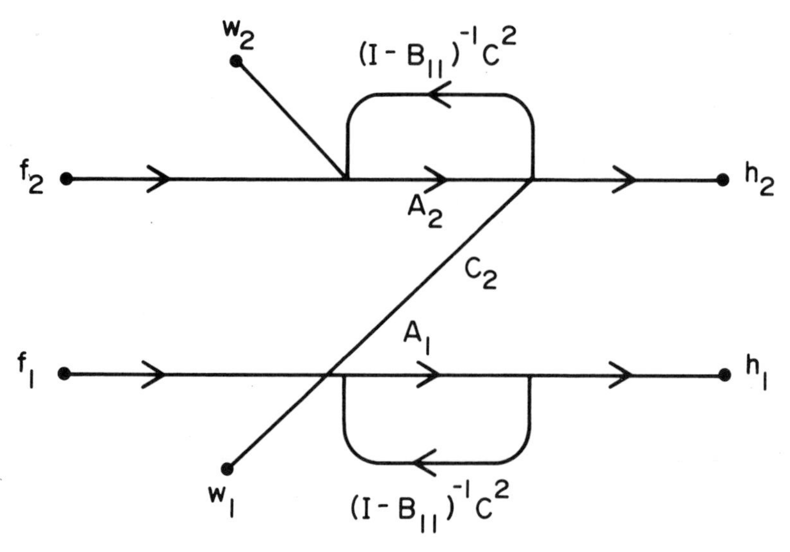

Figure 5.24

Each of these optimization problems is of the form considered in LOP II of Section 2. From the system diagram it is clear that the first component cannot optimize without some knowledge of the behavior of the second component. Consequently, we are led to the extra side condition on the second component. As noted in Section 2, in order for the first controller to solve its local optimization problem it is sufficient to have knowledge of the coordination parameter c. The control structure is depicted in Figure 5.24. The open loop terms w_1 and w_2 depend, respectively, on f_1 and f_2. Both depend on c.

The sensitivity of h_1, h_2, u_1, and u_2 to changes in the coordination parameter c is depicted in Figure 5.25. We now consider various scenarios which can be given for the choice of the coordination parameter c. From a centralized viewpoint, a coordinator may use summary information of the component responses in order to set a value for the coordination parameter. For example, the coordinator may simply observe the motion of the end point of the two link arm and base a decision for the choice of c upon this information. The motion of the end point of the arm is depicted in Figure 5.26 for various values of the coordination parameter.

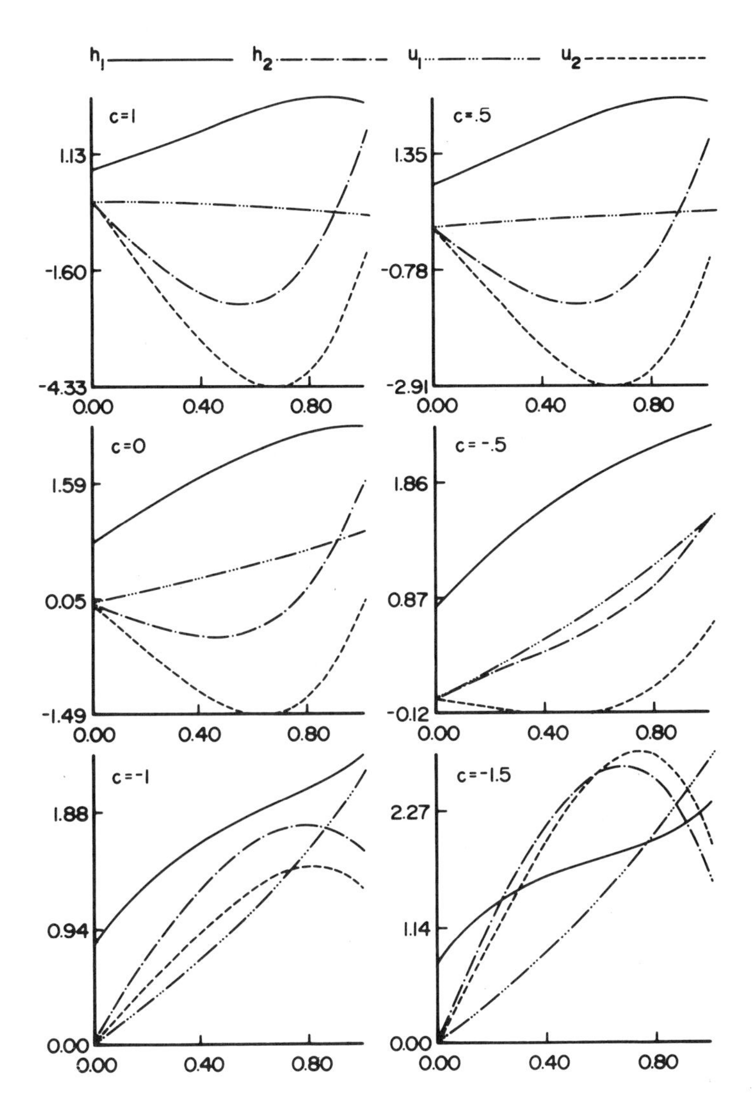

h1
h2
u1
u2
c=1
1.13
-1.60
-4.33
0.00
0.40
0.80
c=.5
1.35
-0.78
-2.91
c=0
1.59
0.05
-1.49
c=-.5
1.86
0.87
-0.12
c=-1
1.88
0.94
0.00
c=-1.5
2.27
1.14

Figure 5.25

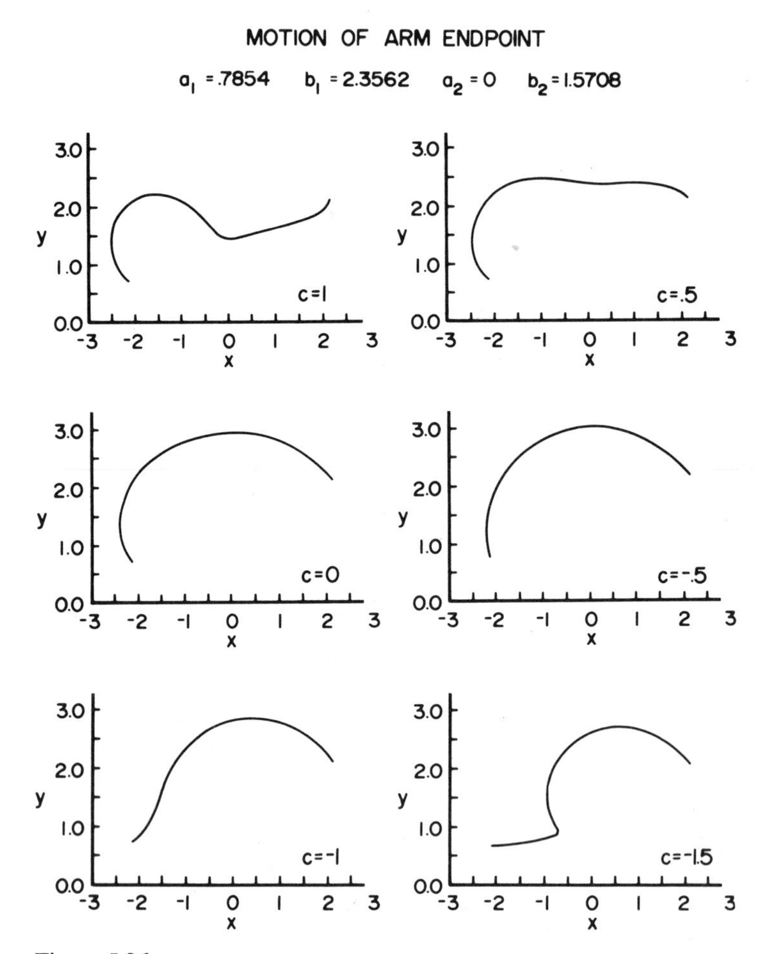

Figure 5.26

Another possibility would be for the coordinator to use information of the dependence of local performance indicies upon the coordination parameter as a basis for making a decision, see Table 5.27.

Alternately, we can envision a command structure set up between the components for the choice of the coordination parameter. Treating c as a design parameter, the second controller could analyze the sensitivity of the solution of its local optimi-

Table 5.27

c	1	.5	0	-.5	-.6	-1
J_1	.31	.34	.74	1.5	1.7	2.7
J_2	40.	20.	7.6	1.3	.9	1.7

zation problem on this parameter, see Figure 5.25. A value of c = -.5 might be chosen, providing a monotone transition to the desired terminal condition. The first controller would then simply solve its local optimization problem using the specified value of c. Similarly, the first component could be allowed to treat c as a design parameter and pass a desirable value for c or a range of values to the second controller.

Summary

Many issues arise in the design of control laws for large scale systems. The primary goal of any design procedure is the effective coordination of system components to achieve the overall control objective. In general problems arise due to restricted information transfer between component controllers and to restricted computational capabilities. That is, individual controllers may not have full information about the overall system upon which to base a control decision or the individual controllers may be unaware of how local control objectives combine to form the overall control objective. In such situations it is natural to ask if the overall control objective can be met while at the same time allowing some level of autonomous local control. Consequently, comparisons of centralized and decentralized decision making, information processing, and computation arise naturally in the design and analysis of control laws for large scale systems.

Methodologies used in the design of control laws for large scale systems are referred to as decentralized control methods, hierarchical control methods, or multilevel methods. In general, some level of local control autonomy, restricted information transfer between components, distributed computation, and a higher level assessment of overall system performance characterize these methods [4, 8, 9, 5].

We have introduced three concepts. The first concept is that of local or component control which requires a surprisingly simple extension of the optimal control methods of Chapter II. The local control problems involve interval constraints as opposed to terminal constraints.

The second concept is that of a stably connected multiple component system. The third concept is the notion of a coordination strategy. Briefly, large scale systems which are stably connected can be coordinated by imposing additional constraints on the components. The individual local control problems are then resolved autonomously.

The two tank decentralized control problem illustrates the coordination of a large scale system with components having hereditary system dynamics. The discussion of system decomposition for finite dimensional state space, time invariant systems serves to point out that many systems are stably connected. Finally, the simple mechanical arm example explores in a preliminary way the problem of enhancing overall system performance using the coordination parameters.

If one views the system components as the basic building blocks then our coordination strategies for stably connected systems can be viewed as a rule for putting together more complex structured systems. From this point of view the open and closed loop control laws arising from the local optimization problems determine how additional components can be "hardwired" into the system.

References

1. H. T. Banks, M. Q. Jacobs and C. E. Langehop, Characterization of the controlled states in $W_2^{(1)}$ of linear hereditary systems, SIAM J. Contr. 13(1975), 611-649.

2. S. B. Black, Coordination of control for large-scale systems, Dissertation, Clemson University, 1983.

3. R. E. Fennell, J. A. Reneke, and S. B. Black, Control coordination for large scale systems, Proc. 1985 American Control Conference, (1985), 590-591.

4. W. Findeisen, F. N. Bailey, M. Byrds, K. Malinowski, P. Tatjewski and A. Wozniak, Control Coordination in Hierarchical Systems, John Wiley and Sons, New York, 1980.

5. D. G. Luenberger, Optimization by Vector Space Methods, John Wiley and Sons, New York, 1969.

6. J. S. Mac Nerney, Hellinger integrals in inner product spaces, J. Elisha Mitchell Sci. Soc., 76(1960), 252-273.

7. A. Manitius and R. Trigginai, Function space controllability of linear retarded systems: A derivation from abstract operator conditions, SIAM J. Contr. Opt., 16(1978) 599-645.

8. M. D. Mesarovic, D. Macko, and Y. Takahara, Theory of Hiearchical Multilevel Systems, Academic Press, New York, 1970.

9. A. N. Michel, On the status of stability of interconnected systems, IEEE Trans. Automat. Contr. AC-28(1983), 639-653.

10. J. A. Reneke, Control of a large scale hereditary system, Proc. Southeastern Symp. on Sys. Theory, Huntsville, Alabama, 1983, 36-39.

11. J. A. Reneke, A strategy for decentralized control of stably connected systems, Mathematical Theory of Networks and Systems, Proc. of the MTNS-83 Inter. Symp., P. A. Fuhrmann, ed., Springer-Verlag, Berlin, 1984.

12. J. A. Reneke and R. E. Fennell, Decentralized control strategies for hereditary systems, Proc. 1985 IEEE Conference on Decision and Control, (1985), 1479-1483.

13. D. L. Russell, Mathematics of Finite Dimensional Control Systems Theory and Design, Marcel Dekker, New York, 1979.

14. D. Salamon, On controllability and observability of time delay systems, IEEE Trans. Automat. Contr., Vol AC-29(1984), 432-438.

15. N. R. Sandell, P. Varaiya, M. Athans and M. G. Safonov, Survey of decentralized control methods for large scale systems, IEEE Trans. Control AC-23(1978), 108-128.

Symbols

Index